T0137117

Springer Proceedings in Mathematics & Statistics

Volume 268

Springer Proceedings in Mathematics & Statistics

This book series features volumes composed of selected contributions from workshops and conferences in all areas of current research in mathematics and statistics, including operation research and optimization. In addition to an overall evaluation of the interest, scientific quality, and timeliness of each proposal at the hands of the publisher, individual contributions are all refereed to the high quality standards of leading journals in the field. Thus, this series provides the research community with well-edited, authoritative reports on developments in the most exciting areas of mathematical and statistical research today.

More information about this series at http://www.springer.com/series/10533

Abdulla Azamov · Leonid Bunimovich
Akhtam Dzhalilov · Hong-Kun Zhang
Editors

Differential Equations and Dynamical Systems

2 USUZCAMP, Urgench, Uzbekistan,
August 8–12, 2017

 Springer

Editors
Abdulla Azamov
Institute of Mathematics
Uzbekistan Academy of Sciences
Tashkent, Uzbekistan

Leonid Bunimovich
School of Mathematics
Georgia Institute of Technology
Atlanta, GA, USA

Akhtam Dzhalilov
Turin Polytechnic University
Tashkent, Uzbekistan

Hong-Kun Zhang
Department of Mathematics and Statistics
University of Massachusetts Amherst
Amherst, MA, USA

ISSN 2194-1009   ISSN 2194-1017   (electronic)
Springer Proceedings in Mathematics & Statistics
ISBN 978-3-030-13180-7   ISBN 978-3-030-01476-6   (eBook)
https://doi.org/10.1007/978-3-030-01476-6

Mathematics Subject Classification (2010): 37E10, 26A18, 28D05, 34C05, 34C28, 34D05, 34D45, 37C55, 37A05, 37A60, 37A50, 37D50, 37D25

This Springer imprint is published by the registered company Springer Nature Switzerland AG
The registered company address is: Gewerbestrasse 11, 6330 Cham, Switzerland

Contents

On Linearization of Circle Diffeomorphisms

Akhadkulov Habibulla, Dzhalilov Akhtam and Konstantin Khanin

Abstract Let f be a circle diffeomorphism with irrational rotation number of bounded type and satisfying a certain Zygmund-type condition depending on a parameter $\gamma > 1$. We prove that f is $C^{1+\omega_\gamma}$—smoothly conjugate to a rigid rotation, where $\omega_\gamma(x) = A|\log x|^{-\gamma+1}$ and $A > 0$. The result completes our resent results in [1].

Keywords Circle diffeomorphisms · Rotation number · Denjoy's inequality · Conjugating map

1 Introduction

The first properties of circle homeomorphisms were studied in a classical work of Poincaré [10]. For an orientation-preserving homeomorphism f of the unit circle $S^1 = \mathbb{R}/\mathbb{Z}$ the limit $\lim_{i \to \infty} L_f^i(x)/i = \rho_f$ exists and does not depend on $x \in \mathbb{R}$, where L_f is a lift of f from S^1 onto $\mathbb{R}$. Here and below L_f^i denotes the ith iteration of L_f. The number $\rho := \rho_f \mod 1$ is called the *rotation number* of f. It is well know that the rotation number is irrational if and only if f has no periodic orbits. Denjoy [3] proved that if f is an orientation-preserving diffeomorphism of the circle with irrational rotation number ρ and $\log f'$ has bounded variation then f

A. Habibulla (✉)
School of Quantitative Sciences, University Utara Malaysia,
CAS 06010, UUM Sintok, Kedah Darul Aman, Malaysia
e-mail: akhadkulov@yahoo.com

D. Akhtam
Turin Polytechnic University, Kichik Halka yuli 17,
Tashkent 100095, Uzbekistan
e-mail: a_dzhalilov@yahoo.com

K. Khanin
Department of Mathematics, University of Toronto, 40 St. George Street,
Toronto, Ontario M5S 2E4, Canada
e-mail: khanin@math.toronto.edu

© Springer Nature Switzerland AG 2018
A. Azamov et al. (eds.), *Differential Equations and Dynamical Systems*,
Springer Proceedings in Mathematics & Statistics 268,
https://doi.org/10.1007/978-3-030-01476-6_1

1

is topological conjugate to the linear rotation $f_\rho : x \to x + \rho$; that is, there exists a homeomorphism h the circle such that $h \circ f \circ h^{-1} = f_\rho$. In the end of 70's the problem of smoothness of the conjugacy of smooth diffeomorphisms was one of the fundamental problems of this direction. The first significant results on smoothness of conjugacy were obtained by Herman in [4]. It was shown that if $f \in C^k$ ($k \geq 3$), its rotation number is irrational and satisfies a certain Diophantine condition then h is in fact $k - 1 - \varepsilon$ times differentiable for any $\varepsilon > 0$, and is analytic if f is analytic. Later Yoccoz [11] extended his results for all Diophantine numbers. In the end of the 80's two different approaches to the Herman's theory were developed by Katznelson and Ornstein [5, 6] and Khanin and Sinai [7, 8]. These approaches gave sharp results on the smoothness of the conjugacy in the case of diffeomorphisms with low smoothness. In 2009, Khanin and Teplinsky [9] developed a conceptually new approach which is entirely based on the idea of cross-ratio distortion estimates. Recently, in [1] we have extended the results of [5–9] for a class of circle diffeomorphisms satisfying a certain Zygmund-type condition as follows. Consider the following one-parameter family of functions: $\Phi_\gamma : [0, 1) \to [0, +\infty)$, $\Phi_\gamma(0) = 0$ and

$$\Phi_\gamma(x) = \frac{x}{(\log \frac{1}{x})^\gamma}, \quad \text{where } 0 < x < 1 \text{ and } \gamma > 0.$$

Denote by $\Delta^2 f'(x, \tau)$ the *second symmetric difference* of f' i.e.,

$$\Delta^2 f'(x, \tau) = f'(x + \tau) + f'(x - \tau) - 2f'(x)$$

where $x \in S^1$ and $\tau \in [0, \frac{1}{2}]$. Suppose that there exists a constant $C > 0$ such that the following inequality holds:

$$\|\Delta^2 f'(\cdot, \tau)\|_{L^\infty(S^1)} \leq C\Phi_\gamma(\tau). \tag{1}$$

Denote by $\mathcal{Z}_\gamma$ the class of circle diffeomorphisms f, whose derivatives f' satisfy (1). The main result of [1] is the following

Theorem 1 *Let $f \in \mathcal{Z}_\gamma$ be a circle diffeomorphism with irrational rotation number ρ.*

(a) *If $\gamma \in (\frac{1}{2}, 1]$ and the partial quotients of ρ satisfies $a_n \leq Cn^\alpha$ for some $\alpha \in (0, \gamma - \frac{1}{2})$ and $C > 0$. Then the conjugating map h between f and f_ρ and its inverse h^{-1} are absolute continuous and h', $(h^{-1})' \in L_2$.*

(b) *If $\gamma > 1$ and the partial quotients of ρ satisfies $a_n \leq Cn^\alpha$ for some $\alpha \in (0, \gamma - 1)$ and $C > 0$. Then the conjugating map h between f and f_ρ and its inverse h^{-1} are C^1 diffeomorphisms.*

In this paper we show that the conjugacy is better than C^1 smooth in the case of $\gamma > 1$ and the rotation number is irrational of bounded type i.e., the partial quotients are bounded. Let $\omega_\gamma(x) = |\log x|^{-\gamma+1}$. Our main result is the following

Theorem 2 *Let $f \in \mathcal{Z}_\gamma$ for some $\gamma > 1$. If the rotation number of f is irrational of bounded type then there exists a constant $A > 0$ such that the conjugating map h between f and f_ρ and its inverse h^{-1} are C^1 diffeomorphisms and*

$$|h'(x) - h'(y)| \le A\omega_\gamma(|x - y|), \quad |(h^{-1}(x))' - (h^{-1}(y))'| \le A\omega_\gamma(|x - y|)$$

for any $x, y \in S^1$, such that $x \neq y$.

2   Preliminaries and Notation

Consider an orientation-preserving circle homeomorphism f with irrational rotation number $\rho := \rho_f$. We shall use the the *continued fraction* expansion for the irrational number $\rho = [a_1, a_2, \ldots, a_n, \ldots]$ which is understood as a limit of the sequence of rational convergents $p_n/q_n = [a_1, a_2, \ldots, a_n]$. The positive integers a_n, $n \ge 1$, are called *partial quotients*. The mutually prime positive integers p_n and q_n satisfy the recurrent relation $p_n = a_n p_{n-1} + p_{n-2}, q_n = a_n q_{n-1} + q_{n-2}$ for $n \ge 1$, where it is convenient to define $p_0 = 0$, $q_0 = 1$ and $p_{-1} = 1$, $q_{-1} = 0$. We take an arbitrary point $x_0 \in S^1$ and fix. Define $\Delta_0^{(n)} := \Delta_0^{(n)}(x_0)$ as the closed interval in S^1 with endpoints x_0 and $x_{q_n} = f^{q_n}(x_0)$, such that, for n odd, x_{q_n} is to the left of x_0, and for n even, it is to its right with respect to the orientation induced from the real line. Denote by $\Delta_i^{(n)} := f^i(\Delta_0^{(n)})$, $i \ge 1$, the iterates of the interval $\Delta_0^{(n)}$ under f. It is well known that the set $\mathbb{P}_n := \mathbb{P}_n(x_0, f)$ of intervals with mutually disjoint interiors defined as

$$\mathbb{P}_n = \{\Delta_i^{(n-1)}, \ 0 \le i < q_n\} \cup \{\Delta_j^{(n)}, \ 0 \le j < q_n\}.$$

determines a partition of the circle for any n. The partition $\mathbb{P}_n$ is called the nth dynamical partition of S^1. Obviously, the partition $\mathbb{P}_{n+1}$ is a refinement of the partition $\mathbb{P}_n$: indeed, the intervals of order n belong to $\mathbb{P}_{n+1}$ and each interval $\Delta_i^{(n-1)}$ $0 \le i < q_n$ is partitioned into $a_{n+1} + 1$ intervals belonging to $\mathbb{P}_n$ such that

$$\Delta_i^{(n-1)} = \Delta_i^{(n+1)} \cup \bigcup_{s=0}^{a_{n+1}-1} \Delta_{i+q_{n-1}+sq_n}^{(n)}. \tag{2}$$

Define $\mathbf{K}_n := \mathbf{K}_n(f) = \max_\xi |\log(f^{q_n}(\xi))'| = \|\log(f^{q_n})'\|_0$. It is well know that if f satisfies the conditions of Denjoy theorem then $\mathbf{K}_n \le v$ where $v = Var_{S^1} \log f'$ (see [8]). This inequality is know as Denjoy's inequality and it has very important applications in the theory of circle homeomorphisms. Using this, it can be shown that the intervals of the dynamical partition $\mathbb{P}_n$ have exponentially small length. Indeed, one finds the following

Theorem 3 *Let f be an orientation-preserving diffeomorphism of the circle with irrational rotation number and $\log f'$ has bounded variation. There exist constants $C_0 > 1$ and $0 < \mu < \lambda < 1$ depending only on f such that*

(i) *For any $\Delta^{(n)} \in \mathbb{P}_n$ we have $|\Delta^{(n)}| \leq C_0 \lambda^n$;*

(ii) *If the rotation number of f is of bounded type then for any two adjacent intervals $\Delta^{(n)}$ and $\tilde{\Delta}^{(n)}$ of $\mathbb{P}_n$ we have $C^{-1}|\tilde{\Delta}^{(n)}| < |\Delta^{(n)}| < C|\tilde{\Delta}^{(n)}|$;*

(iii) *If the rotation number of f is of bounded type then for any $\Delta^{(n)} \in \mathbb{P}_n$ we have $|\Delta^{(n)}| \geq C_0^{-1} \mu^n$.*

The proof of the item (i) of this can be found in [1] and the proofs of the items (ii) and (iii) can be found [2].

3 Some Supporting Lemmas

Denote $\widehat{\Delta}_0^{(n)} = \Delta_0^{(n)} \cup \Delta_0^{(n-1)}$. Let $i_n : S^1 \to \mathbb{N}_0$ be the first entrance time of x in $\widehat{\Delta}_0^{(n)}$; that is, $i_n(x) = \min\{i \geq 0 : f^i(x) \in \widehat{\Delta}_0^{(n)}\}$. Define $\zeta_n : S^1 \to \mathbb{R}$ as follows

$$\zeta_n(x) = \sum_{s=0}^{i_n(x)-1} \log f'(f^s(x)).$$

The following lemmas are needed to derive our main result.

Lemma 1 *Let f satisfies the conditions of Theorem 2. Then ζ_n is a Cauchy sequence.*

Proof By the definition of i_n :

$$i_n(x) = \begin{cases} 0, & \text{if } x \in \widehat{\Delta}_0^{(n)} \\ q_{n-1} - j, & \text{if } x \in \Delta_j^{(n)} \\ q_n - i, & \text{if } x \in \Delta_i^{(n-1)} \end{cases}$$

where $0 < j < q_{n-1}$ and $0 < i < q_n$. This and by (2) we get

$$i_{n+1}(x) = \begin{cases} 0, & \text{if } x \in \widehat{\Delta}_0^{(n+1)} \\ q_{n+1} - j, & \text{if } x \in \Delta_j^{(n)} \\ q_{n+1} - (i + q_{n-1} + sq_n), & \text{if } x \in \Delta_{i+q_{n-1}+sq_n}^{(n)} \\ q_n - i, & \text{if } x \in \Delta_i^{(n+1)} \end{cases}$$

where $0 < j < q_{n-1}$, $0 < i < q_n$ and $0 \leq s < a_{n+1}$. Therefore

$$i_{n+1}(x) - i_n(x) = \begin{cases} 0, & \text{if } x \in \widehat{\Delta}_0^{(n)} \cup \Delta_i^{(n+1)} \\ a_{n+1}q_n, & \text{if } x \in \Delta_j^{(n)} \\ (a_{n+1} - s - 1)q_n, & \text{if } x \in \Delta_{i+q_{n-1}+sq_n}^{(n)} \end{cases}$$

where $0 < j < q_{n-1}$, $0 < i < q_n$ and $0 \leq s < a_{n+1}$. Using the last relation we get

$$\|\zeta_{n+1}(x) - \zeta_n(x)\|_\infty \le a_{n+1} \mathbf{K}_n. \tag{3}$$

By Theorem 7.1 and Lemma 8.2 in [1] we have $\mathbf{K}_n \le Cn^{-\gamma}$. Since the rotation number is bounded type we get

$$\|\zeta_{n+p}(x) - \zeta_n(x)\|_\infty \le C \sum_{m=n}^{n+p-1} \frac{1}{m^\gamma}. \tag{4}$$

We conclude from (4) that ζ_n is a Cauchy sequence.

Let $\zeta(x) = \lim_{n \to \infty} \zeta_n(x)$.

Lemma 2 *Let f satisfies the conditions of Theorem 2. Then $\zeta : S^1 \to \mathbb{R}$ is continuous and satisfies the relation*

$$\zeta(f(x)) = \zeta(x) - \log f'(x). \tag{5}$$

Proof It is easy to see that for any $x \in S^1$ there exists $n_0 := n_0(x)$ such that $i_n(f(x)) = i_n(x) - 1$ for all $n \ge n_0$. This and by the definition of ζ_n we get

$$\zeta_n(f(x)) = \zeta_n(x) - \log f'(x)$$

for all $n \ge n_0$. Taking the limit as $n \to \infty$ we get (5). Next we show ζ is continuous at $x = x_0$. One can see $\zeta_n(x_0) = 0$ for all $n \ge 1$, so $\zeta(x_0) = 0$. Take any $z \in \widehat{\Delta}_0^{(n)}$. It is obvious that $i_j(z) = 0$ for every $j \le n$, so $\zeta_j(z) = 0$ for every $j \le n$. In particular

$$\zeta_{n+p}(z) = \sum_{m=0}^{p-1} \zeta_{n+m+1}(z) - \zeta_{n+m}(z).$$

This and the relation (4) imply

$$|\zeta_{n+p}(z)| \le C \sum_{m=n}^{n+p-1} \frac{1}{m^\gamma}.$$

Consequently

$$\lim_{n \to \infty} \sup_{z \in \widehat{\Delta}_0^{(n)}} |\zeta(z)| = 0.$$

Hence ζ is continuous at $x = x_0$. Denote by $\Xi = \{x_i := f^i(x); \ i \in \mathbb{N}\}$ the trajectory of x_0. Since ζ is continuous at $x = x_0$ and $\log f'$ is continuous on S^1, by (5) it implies that ζ is continuous on Ξ. Note that $i_n : S^1 \to \mathbb{R}$ is continuous in the interior of each element of the partition $\mathbb{P}_n$ for every $n \ge 1$. As a consequence ζ_n is continuous in the interior of each element of the partition $\mathbb{P}_n$ for every $n \ge 1$. Thus the limit function ζ is continuous on $x \in S^1 \setminus \Xi$.

Lemma 3 *Let f satisfies the conditions of Theorem 2. There exists $C > 0$ such that*

$$|\zeta(x) - \zeta(y)| \leq C\omega_\gamma(|x - y|) \tag{6}$$

for any $x, y \in S^1$, such that $x \neq y$.

Proof Consider the points x_i and $x_{i+q_{n-1}+sq_n}$ where $1 \leq s \leq a_{n+1}$. It is clear that $x_i, x_{i+q_{n-1}+sq_n} \in \Delta_i^{(n-1)}$. The relation (5) implies

$$|\zeta(x_{i+q_{n-1}+sq_n}) - \zeta(x_i)| \leq a_{n+1}\mathbf{K}_n.$$

Consequently, for any $x_j \in \Xi \cap (\Delta_i^{(n-1)} \setminus \Delta_i^{(n+1)})$ we have

$$|\zeta(x_j) - \zeta(x_i)| \leq \sum_{m=n}^{\infty} a_{m+1}\mathbf{K}_m.$$

Since a_{m+1} is bounded we get

$$|\zeta(x_j) - \zeta(x_i)| \leq C\sum_{m=n}^{\infty} \frac{1}{m^\gamma} \leq \frac{C}{n^{\gamma-1}}. \tag{7}$$

It is obvious that

$$|\Delta_i^{(n+1)}| \leq |x_j - x_i| \leq |\Delta_i^{(n-1)}|.$$

This and Theorem 3 imply

$$n = \mathcal{O}\left(\frac{1}{|\log|x_j - x_i||}\right). \tag{8}$$

Combining (7) with (8) we can assert that

$$|\zeta(x_j) - \zeta(x_i)| \leq \frac{C}{|\log|x_j - x_i||^{\gamma-1}}. \tag{9}$$

Since Ξ is dense in S^1, the function ζ can be continuously extended to the whole of S^1 verifying the inequality (9).

4   Proof of Theorem 2

Consider the function $\varphi : S^1 \to \mathbb{R}$ defined as

$$\varphi(x) = e^{\zeta(x)} \left(\int_{S^1} e^{\zeta(t)} dt \right)^{-1}.$$

It is clear φ is continuous and positive on S^1. We claim that φ satisfies the *homological equation*

$$\varphi(f(x)) = \frac{1}{f'(x)} \varphi(x), \quad x \in S^1. \tag{10}$$

Indeed, by the inequality (5) we get

$$\varphi(f(x)) = e^{\zeta(f(x))} \left(\int_{S^1} e^{\zeta(t)} dt \right)^{-1} = e^{\zeta(x) - \log f'(x)} \left(\int_{S^1} e^{\zeta(t)} dt \right)^{-1} = \frac{1}{f'(x)} \varphi(x).$$

Next we show that the C^1—smooth diffeomorphism

$$h(x) = \int_{x_0}^{x} \varphi(t) dt, \quad x \in S^1$$

conjugates f and f_ρ. Using the relation (10) we get

$$h(f(x)) = h(x) + \int_{x_0}^{f(x_0)} \varphi(t) dt. \tag{11}$$

Denote by H and F the lift functions of h and f respectively. From the relation (11) it follows that

$$H(F^n(x)) = H(x) + n \int_{x_0}^{f(x_0)} \varphi(t) dt, \quad x \in \mathbb{R} \tag{12}$$

for all $n \geq 1$. It is well known (see for instance [4]) that there exists a one periodic functions $\widetilde{H}$ such that $H = \widetilde{H} + \mathrm{Id}$. Therefore, by (12) we get

$$\frac{F^n(x) - x}{n} = \frac{\widetilde{H}(x) - \widetilde{H}(F^n(x))}{n} + \int_{x_0}^{f(x_0)} \varphi(t) dt. \tag{13}$$

Taking the limit as $n \to \infty$ we get

$$\rho = \int_{x_0}^{f(x_0)} \varphi(t) dt.$$

Hence $h \circ f = f_\rho \circ h$. From Lemma 3 it follows that

$$|\varphi(x) - \varphi(y)| \le C\omega_\gamma(|x - y|)$$

and consequently

$$|h'(x) - h'(y)| \le C\omega_\gamma(|x - y|)$$

for any $x, y \in S^1$, such that $x \ne y$. Since h and h^{-1} are diffeomorphisms we can easily show that

$$|(h^{-1}(x))' - (h^{-1}(y))'| \le C\omega_\gamma(|x - y|)$$

for any $x, y \in S^1$, such that $x \ne y$. Thus, Theorem 2 is completely proved.

References

1. Akhadkulov, H., Dzhalilov, A., Khanin, K.: Notes on a theorem of Katznelson and Ornstein. Dis. Con. Dyn. Sys. **37**(9), 4587–4609 (2017)
2. Akhadkulov, H., Dzhalilov, A., Noorani, M.S.: On conjugacies between piecewise-smooth circle maps. Nonlinear Anal. Theory Methods Appl. **99**, 1–15 (2014)
3. Denjoy, A.: Sur les courbes définies par les équations différentielles à la surface du tore. J. Math. Pures Appl. **11**, 333–375 (1932)
4. Herman, M.: Sur la conjugaison différentielle des difféomorphismes du cercle à des rotations. Inst. Hautes Etudes Sci. Publ. Math. **49**, 5–234 (1979)
5. Katznelson, Y., Ornstein, D.: The differentiability of the conjugation of certain diffeomorphisms of the circle. Ergod. Theor. Dyn. Syst. **9**, 643–680 (1989)
6. Katznelson, Y., Ornstein, D.: The absolute continuity of the conjugation of certain diffeomorphisms of the circle. Ergod. Theor. Dyn. Syst. **9**, 681–690 (1989)
7. Khanin, K.M., Sinai, Y.G.: A new proof of M. Herman's theorem. Commun. Math. Phys. **112**, 89–101 (1987)
8. Khanin, K.M., Sinai, Y.G.: Smoothness of conjugacies of diffeomorphisms of the circle with rotations. Russ. Math. Surv. **44**, 69–99 (1989); Trans. Usp. Mat. Nauk. **44**, 57–82 (1989)
9. Khanin, K.M., Teplinsky, AYu.: Herman's theory revisited. Invent. Math. **178**, 333–344 (2009)
10. Poincaré, H.: Mémoire sur les courbes définies par une équation différentielle (I). J. Math. Pures Appl. **7**, 375–422 (1881)
11. Yoccoz, J.C.: Conjugaison différentiable des difféomorphismes du cercle dont le nombre de rotation vérifie une condition diophantienne. Ann. Sci. École Norm. Sup. (4) **17**(3), 333–359 (1984)

The Fujita and Secondary Type Critical Exponents in Nonlinear Parabolic Equations and Systems

Aripov Mersaid

Abstract In this work, demonstrated the possibilities of the self-similar approach to the studying of qualitative properties of nonlinear reaction diffusion equation and system such as finite speed of a perturbation, Fujita and secondary type critical exponents of a global solvability. Asymptotic of the self-similar solutions in a secondary critical case is established. Based on the computer modeling of nonlinear processes described by nonlinear degenerate parabolic equation and cross system discussed. The problem choosing an initial approximation for numerical solution depending on a value of numerical parameters is solved.

Keywords Parabolic · Degenerate · Equation · System · Self-similar · Fujita

1 Introduction

This paper devoted to a various extensions of a result of Fujita [1] and secondary critical exponent for the initial value problem to the reaction-diffusion equation

$$\frac{\partial u}{\partial t} = \nabla(u^{m-1}|\nabla u^k|^{p-2}\nabla u) + div(c(t)u) + \gamma(t)u^\beta, \tag{1}$$

$$u(0, x) = u_0(x) \geqslant 0, \quad x \in R^N, \tag{2}$$

in $Q = (t > 0, x \in R^N)$ with $m, \beta, k \geqslant 1, p \geqslant 2, \nabla(\cdot) = grad_x(\cdot), 0 < c(t),$ $\gamma(t) \in C(0, \infty).$

A. Mersaid
National university of Uzbekistan, Tashkent, Uzbekistan
e-mail: mirsaidaripov@mail.ru

A. Azamov et al. (eds.), *Differential Equations and Dynamical Systems*,
Springer Proceedings in Mathematics & Statistics 268,
https://doi.org/10.1007/978-3-030-01476-6_2

Also to the nonlinear degenerate parabolic cross system (see [2–17])

$$A(u, v) = -\frac{\partial u}{\partial t} + div\left(v^{m_1-1}\left|\nabla u^k\right|^{p-2}\nabla u\right) - div(c(t)u) + \gamma(t)u^{\beta_1} = 0,$$

$$B(u, v) = -\frac{\partial v}{\partial t} + div\left(u^{m_2-1}\left|\nabla v^k\right|^{p-2}\nabla v\right) - div(c(t)v) + \gamma(t)v^{\beta_2} = 0, \quad (3)$$

$$u\,(0, x) = u_0\,(x) \geqslant 0, \quad v\,(0, x) = v_0(x) \geqslant 0, \quad x \in R^N, \tag{4}$$

where $m_i, \beta_i \in R, \ i = 1, 2, \ p \geqslant 2, \ k \geqslant 1$ are the given numerical parameters, $\nabla(\cdot) = grad_x(\cdot)$.

The problem (1), (2) describes many nonlinear processes, for instance the processes of nonlinear filtration in liquid and gas, the thermal conductivity, nonlinear reaction diffusion, when the thermal conductivity coefficient is a power function of the derivative in the presence of a convective transfer with speed $c(t)$ and source [2, 3, 18–23]. The Eq. (1) is a base for modeling of the many physical processes too [1, 17–22, 24–26].

Notice that the class of Eq. (1) contains the linear diffusion equation, ($p = 2, \ m = 1$), commonly known as the heat equation, $\partial_t u = \Delta u$; the nonlinear diffusion equation $\partial_t u = \Delta u^k$, known as the porous medium equation ($p = 2, \ m > 1$), or the fast diffusion equation ($p = 2, \ m < 1$), and the gradient-dependent diffusion equation, $\partial_t u = \text{div}(|\nabla u|^{p-2}\nabla u) := \Delta_p u$, that is, the $p-$Laplacian equation ($p \neq 2, \ m = 1$). When $p \neq 2$ and $m \neq 1$, Eq. (1) is called the doubly nonlinear diffusion equation, due to the fact that its diffusion term depends nonlinearly on both the unknown density u, and its gradient ∇u. Such gradient-dependent diffusion equations appear in several models in non-Newtonian fluids [16], in glaciology [13, 23], and in turbulent flows in porous media [17]. For more details on these models, we refer to the work [2–10, 22–24], and the references therein.

Equation (1) is good combination for of slowly diffusion ($k(p - 2) + m - 1 > 0$), fast diffusion $k(p - 2) + m - 1 < 0$ and other diffusion cases too. One of the particular features of problem (1) is that the equation is degenerate at points where $u = 0$ or $\nabla u = 0$. Hence, there is no classical solution in general. Therefore, we consider weak solution with property

$$0 \leqslant u(t, x), \quad u^{m-1}\left|\nabla u^k\right|^{p-2}\nabla u \in C(Q),$$

satisfying to Eq. (1) in tense of distribution [24].

System (3) describes the processes of reaction—diffusion, heat conductivity, polytrophic filtration of gas and liquid ($k = 1, \ p = 2$), biological population and etc. in the two componential nonlinear medium with source, and convective transfer speed of which $c(t)$ depends on time. A specifically properties of this equation and system is its degenerating. Therefore, we need to investigate the weak solution, because in this case solutions of problem (3), (4) may do not exist in the classical tense.

For system (3) in general we will study a class of weak solutions with properties

$$0 \leqslant u(t, x), \quad (u^{m_1-1}(t, x)|\nabla u^k|^{p-2}\nabla u) \in C(Q),$$

$$0 \leqslant v(t, x), \quad (v^{m_2-1}(t, x)|\nabla v^k|^{p-2}\nabla v) \in C(Q),$$

and satisfying to (1) in the sense of a distribution [24].

Analytical solving of the considered nonlinear problem is very complicated. Therefore, now computing experiment becomes almost unique means for solving of the nonlinear problems. But, before a numerical computing are required investigation of qualitative properties of different type solutions arising on depended of value of numerical parameters of the considered problem necessary to study the qualitative properties of solutions.

One of effective method for investigation qualitative properties of considered problem is self-similar, approximately self-similar approach [1, 18–20, 24]. For this goal, we use method of nonlinear splitting algorithm [20], which allowed constricting the system of self-similar equation for (1) and system (2). This approach intensively used by many authors [18–20, 24] for investigation of the new properties of solution such as finite speed of perturbation, blow up properties, localization of solutions and so on [24].

2 Fujita Type Global Solvability

2.1 Case of Single Equation

Consider a global solvability of the problem (1), (2). Different qualitative properties of solution for the particular value of numerical parameters of the Cauchy and boundary value problem to the Eq. (1) intensively studied by many authors [1–8, 18–24]. First Fujita [1] for the problem (1) showed that if $\gamma(t) = 1$, $c(t) = 0$, $m = 1$, $p = 2$, $1 < \beta \leqslant 1 + 2/N$, all solutions are blow up in time [1], while if $\beta > 1 + 2/N$ the problem has a global solution for small initial data. Value of numerical parameter when $\beta = 1 + 2/N$ is called the Fujita type critical exponent.

Samarskii A.A. and etc. [24] showed that condition of the global solvability when $\gamma(t) = 1$, $c(t) = 0$, $p = 2$ is $\beta > m + 1 + 2/N$. After V. Galaktionov establish the following condition of the global solvability $\beta > p - 1 + p/N$ (see [24]) when in (1) $\gamma(t) = 1$, $c(t) = 0$, $m = 1$, $k = 1$ (p–Laplacian equation). More general condition of a global solvability when $c(t) = 0$, $k = 1$ were established in [19], when $\gamma(t) = 1$, $c(t) = 0, k = 1$ the variable density case of the Eq. (1) considered in works [2–5, 20–23]. The condition of the global solvability in the case $c(t) = 0, k = 1$ obtained in the work [20] The role of the Fujita and secondary critical exponents type intensively discussed in literature [1–6, 18–24].

Usually for establishing blow up properties solution applied the. In [5] authors to show the blow up phenomena not use technique the Zel'dovich–Kompaneets–Barenblatt solutions [24], since the construction of such type of function is more complicated for considered problem. Therefore, authors obtain a result by multiplying on a special factor, which has convenient properties. In particular, by choosing the parameters of the factor and using the properties of the solution, obtained the inequality, which allows proving blow up property. Considering semi linear case of the system (3) when in $k = m = 1$, $p = 2$ first Escobedo-Herero [11] establish the Fujita type global solvability. Notice that the Fujita type global solvability of the problem (3), (4) is not studied yet.

This paper discusses problem the Fujita type global solvability and secondary critical exponents for double nonlinear degenerate equation (1) and system (3) using self-similar approach [20]. The algorithm establishing both critical exponents using self-similar analysis of solutions is suggested. Based on an invariant group (self-similar) analysis the method of establishing of a value of the Fujita type critical exponents for single degenerate type parabolic equation and system (3) is given. The Fujita type condition of a global solvability to the problem (1), (2) are established. It is shown that formally a value of the second critical exponent for degenerate type double nonlinear parabolic equation and they system is the roots of the linear algebraic system equations. The estimate of weak solutions to the problem (3), (4) is obtained. Depending on value of numerical parameters, the problem of an appropriate initial approximation solution for an iterative process, leading to the quick convergence with necessary accuracy is solved.

In recent years, as mentioned above many authors [2–10, 18, 19, 21–23] have studied the different qualitative properties of solutions to the Cauchy problem (1) and their variants (see ([2–10, 23] and the references therein)). Zheng et al. [22] investigate the blow-up properties of the positive solution of the Cauchy problem (1) in the case $c(t) = 0$, $\gamma(t) = 1$, and established a secondary critical exponent for the decay initial value at infinity. They notice in this case the problem of the existence and nonexistence of global solutions of the Cauchy problem not considered.

Under some suitable assumptions, the existence, uniqueness and regularity of a weak solution to the Cauchy problem (1) and their variants have been extensively investigated by many authors (see [2–11, 20–23] and the references therein).

The first goal of this paper is to study the blow-up behavior of solution $u(x, t)$ of (1) when the initial data $u_0(x)$ has slow decay near $x = \infty$. For instance, in the following case

$$u_0(x) \cong M|x|^{-a}, \quad M > 0, \quad a \geq 0, \tag{5}$$

In recent years, many authors have studied the properties of solutions to the Cauchy problem (1), (2) and their variants [2–10, 23] and the references therein). In particular, J.-S. Guo and Y. Y. Guo (see [22] and references) obtained the secondary critical exponent for the case $k = 1$, $p = 2$ and shows there exists a secondary critical exponent $a^* = 2/(p - m)$ such that the solution $u(x, t)$ of (1) blows up in finite time for the initial data $u_0(x)$, which behaves like $|x|^{-a}$ at $x = \infty$ if $a \in (0, a^*)$, and there exists a global solution for the initial data $u_0(x)$, which behaves like $|x|^{-a}$ at $x = \infty$ if $a \in (a^*, N)$.

Mu et al. [22] studied the secondary critical exponent for the p-Laplacian equation ($m = 1$) with slow decay initial values and shows that there exists a secondary critical exponent $a_c^* = (p/(q + 1 - p))$ such that the solution $u(x, t)$ of (1) blows up in finite time for the initial data $u_0(x)$ which behaves like $|x|^{-a}$ at $x = \infty$ if $a \in (a_c^*, N)$, and there exists a global solution for the initial data $u_0(x)$, which behaves like $|x|^{-a}$ at $x \to \infty$ if $a \in (a_c^*, N)$.

Recently, Zheng and Mu [9] also investigated the secondary critical exponent for the doubly degenerate parabolic equation with slow decay initial values and obtained similar results. Introduce the function

$$z_+(t, x) = \bar{u}(t)\bar{f}(\xi), \quad \bar{u}(t) = [T + (\beta - 1)\int_0^t \gamma(y)dy]^{-\frac{1}{\beta-1}},$$

$$f(\xi) = (a - b\xi^\gamma)^{\gamma_1}, \quad a > 0. \tag{6}$$

$$b = (k(p - 2) + m - 1)p^{-p/(p-1)}, \quad \gamma = \frac{p}{p-1}, \quad \gamma_1 = \frac{p-1}{k(p-2) + m - 1}.$$

Below considering the problem Cauchy (1), (2); (3), (4) the algorithm for construction of the Fujita type a critical exponent is suggested and has establish the Fujita type for critical exponent. Applying this algorithm, condition of a global solvability Cauchy problem (1), (2) and (3), (4) are obtained.

3 Main Results

3.1 Fujita Type Critical Exponent to the Problem (1)

Theorem 1 *Assume* $k(p - 2) + m - 1 > 0$,

$$\gamma(t)\tau(t)[\bar{u}(t)]^{\beta-[k(p-2)+m-1]} < N/p, \quad t > 0, \quad u_0(x) \le z_+(0, x), \quad x \in R^N, \tag{7}$$

then $u(t, x) \le z_+(t, x)$ *in* Q.

For the problem (1) the Fujita type critical exponent is

$$\gamma(t)\tau(t)[\bar{u}(t)]^{\beta-[k(p-2)+m-1]} = N/p, \quad t > 0.$$

This result consist all early known results other authors (Fujita, Samarskii A.A., Kurdyumov S.P., Galaktionov V.A., Mikhaylov A.P. and others) on a global solvability problem Cauchy (1), (2). In the case $c(t) = 0$, $\gamma(t) = 1$ we obtain all early known Fujita type condition of a global solvability [1, 18–22, 24]

$$\beta > k(p - 2) + m + p/N.$$

Value of the Fujita type critical exponent is $\beta = \beta_* = k(p - 2) + m + p/N$.

Corollary 1 *In the critical case the Eq. (1) always is a self-similar*

$$\xi^{1-N} \frac{d}{d\xi} \left(\xi^{N-1} f^{m-1} \left| \frac{df^k}{d\xi} \right|^{p-2} \frac{df}{d\xi} \right) + \frac{\xi}{p} \frac{df}{d\xi} + (N/p)(f^\beta + f) = 0.$$

Theorem 2 *Let us $1 < \beta \leq k(p - 2) + m + p/N$. Then all solutions of the problem (1), (2) are blow up in time for $u_0(x) \neq 0$, $x \in R^N$.*

3.2 The Fujita Type Global Solvability for the System (3)

Consider the functions

$$u_+(t, x) = \bar{u}(t)\bar{f}(\xi), \quad v_+(t, x) = \bar{v}(t)\,\bar{\psi}(\xi),$$

$$\xi = |\eta|/[\tau(t)]^{1/p}, \quad \tau(t) = a_1(T + t)^{1/a_1}, \quad \eta = \int_0^t c(y)dy - x, \quad x \in R^N$$

$$a_1 = (\beta_1 - 1)(\beta_2 - 1)/(\beta_1 - 1)(\beta_2 - 1) - (m_1 - 1)(\beta_1 - 1) - (\beta_2 - 1)(p - 2),$$

$$\bar{f}(\xi) = (a - \xi^\gamma)_+^{q_1}, \quad \bar{\psi}(\xi) = (a - \xi^\gamma)_+^{q_2}, \quad a > 0, \ \gamma = p/(p - 1)$$

$$q_1 = \frac{(p - 1)(k(p - 2) - (m_1 - 1))}{q}, \quad q_2 = \frac{(p - 1)(k(p - 2) - (m_2 - 1))}{q},$$

$$q = [k(p - 2)]^2 - (m_1 - 1)(m_2 - 1)$$

Theorem 3 *Assume*

$$(\beta_1 - 1)(\beta_2 - 1) - (m_1 - 1)(\beta_1 - 1) - (\beta_2 - 1)k(p - 2) > 0,$$

$$\frac{\beta_2 - 1}{(\beta_1 - 1)(\beta_2 - 1) - (m_2 - 1)(\beta_2 - 1) - k(p - 2)(\beta_1 - 1)} < N/p,$$

$$\frac{\beta_1 - 1}{(\beta_1 - 1)(\beta_2 - 1) - (m_1 - 1)(\beta_1 - 1) - k(p - 2)(\beta_2 - 1)} < N/p,$$

$$u_0(x) \leqslant u_+(0, x), \quad v_0(x) \leqslant v_+(0, x), \quad x \in R^N.$$

Then for the solution of the problem (1), (2) in Q the estimate

$$u(t, x) \leqslant u_+(t, x), \quad v(t, x) \leqslant v_+(t, x). \tag{8}$$

is hold.

Proof of the Theorem 1. Consider the following self-similar solution of the Eq. (1)

$$u(t, x) = \bar{u}(t) f(\xi), \quad \xi = |\eta| \tau^{-1/p}, \quad \eta = \int_0^t c(y) dy - x,$$

$$\tau(t) = \bar{u}^{\beta - [k(p-2)+m]} / [\beta - (k(p-2)+m)],$$

where the function

$$\bar{u}(t) = [T + (\beta - 1) \int_0^t \gamma(y) dy]^{-\frac{1}{\beta - 1}}$$

is solution of the equation

$$\frac{d\bar{u}}{dt} = -\gamma(t) \bar{u}^{\beta},$$

$f(\xi)$ satisfy to an approximately self-similar equation

$$\xi^{1-N} \frac{d}{d\xi} \left(\xi^{N-1} f^{m-1} \left| \frac{df^k}{d\xi} \right|^{p-2} \frac{df}{d\xi} \right) + \frac{\xi}{p} \frac{df}{d\xi} + s(t)(f^{\beta} + f) = 0, \tag{9}$$

where $s(t) = \gamma(t)\tau(t)[\bar{u}(t)]^{\beta - [k(p-2)+m]}$.

In particular when $\gamma(t) = 1$ from (7) we have a self-similar equation

$$A(f) \equiv \xi^{1-N} \frac{d}{d\xi} \left(\xi^{N-1} f^{m-1} \left| \frac{df^k}{d\xi} \right|^{p-2} \frac{df}{d\xi} \right) + \frac{\xi}{p} \frac{df}{d\xi} + d(f^{\beta} + f) = 0, \tag{10}$$

where $d = \frac{1}{\beta - [k(p-2)+m]}$.

Easy to check that for the function $\overline{f}(\xi)$ after simple calculation we have

$$\xi^{1-N} \frac{d}{d\xi} \left(\xi^{N-1} \overline{f}^{m-1} \left| \frac{d\overline{f}^k}{d\xi} \right|^{p-2} \frac{d\overline{f}}{d\xi} \right) + \frac{\xi}{p} \frac{d\overline{f}}{d\xi} = -(N/p)\overline{f}(\xi).$$

Therefore from (8) we have

$$A(\overline{f}) = [[-(N/p) + \gamma(t)\tau(t)\bar{u}(t)]^{\beta - [k(p-2)+m]} + \gamma(t)\tau(t)[\bar{u}(t)]^{\beta - [k(p-2)+m]} \overline{f}^{\beta - 1}] \overline{f}.$$

According condition of the theorem for small value of a we have

$$A(\overline{f}) \le 0 \quad \text{in} \quad \xi < a^{(p-1)/p}.$$

Therefore, according the comparison principle conclude

$$f \le \overline{f} \quad \text{in} \quad \xi < a^{(p-1)/p}.$$

It means that

$$u(t, x) \le u_+(t, x) = \overline{u}(t)\overline{f}(\xi), \quad in \quad Q.$$

Proof of the Theorem 1 completed.

4   The Second Critical Exponent Case

Recently Zheng, Chunlai Mu, Dengming Liu, Xianzhong Yao, and Shouming Zhou for the decaying initial data establish a secondary critical exponent to the problem (1), (2) when $\gamma(t) = 1$. They for the case $c(t) = 0$, $\gamma(t) = 1$ established that if $u_0(x) \approx M|x|^{-a}$, $M > 0$ then value $a = a_* = p/(\beta - k(p - 2) + m)$ is secondary critical exponent for the problem Cauchy. The cases $k = 1$, $\gamma(t) = 1$, $c(t) = 0$, $\gamma(t) = 1$, $c(t) = 0$, $k = 1$, $p = 2$ considered in works [1] In particular, J.S. Guo and Y.Y. Guo (see [22]) when $c(t) = 0$, $\gamma(t) = 1$, $k = 1$, $p = 2$ obtained the secondary critical exponent for the porous medium type equation in high dimensions and proved existing a secondary critical exponent $a = a_* = 2/(\beta - m)$ such that if $u_0(x) \approx |x|^{-a}$ the solution of (1) blows up in finite time for the initial data, which behaves like $|x|^{-a}$ at ∞ if a belongs to $(0, a_*)$, and there exists a global solution if a belongs to (a_*, N).

Below we establish asymptotic behavior of the solutions in the secondary critical exponent case.

Introduce the function

$$\overline{f}(\xi) = (a + \xi^{\gamma})^{\gamma_1}, \quad \gamma = \frac{p}{p-1}, \quad \gamma_1 = -\frac{p-1}{\beta - (k(p-2) + m)}.$$

Theorem 4 *Let us* $\beta > \max(k(p-2) + m), [(k(p-2) + m)N]/(N-p), p < N$, *then the regular vanishes at infinity solutions of the equation (1) has an asymptotic representation*

$$f(\xi) = c(m, l, p, k, N, \beta)(a + \xi^{\frac{p}{p-1}})^{-\frac{p-1}{\beta-(k(p-2)+m)}}(1 + o(1)), \tag{11}$$

where

$$c(m, p, k, N, \beta) = \left(|k\gamma_1|^{p-2}(p-1)\frac{(N-p)\beta - (k(p-2)+m)N}{\beta - (k(p-2)+m)}\right).$$

In the work [24] in the case $p = 2$, $k = l = 1$, $\gamma(t) = 1$, $c(t) = 0$ the following formal asymptotic of solution is given

$$f(\xi) \approx c\xi^{-\frac{2}{\beta-m}}$$

which used for numerical solution. But, value of constants c is not known. We notice according the Theorem 1 value of constants c is

$$c = \left[\frac{(N-2)\beta - Nm}{\beta - m}\right]^{\frac{1}{\beta-m}}, \quad \beta > N/N - 2, \ N \geqslant 3.$$

Mentioned authors using this asymptotic of solution solves numerically. But without out proving of the Theorem 1 and finding value of constant c. Consider particular case of the Eq. (1) when $\gamma(t) = 1$, $c(t) = 0$.

Then notice that from (9) in the case $m = 1$, $p = 2$, $k = l = 1$ we have

$$c(1, 1, 2, 1, N, \beta) = \left[\frac{(N-2)\beta - N}{\beta - 1}\right]^{\frac{1}{\beta-1}}.$$

In particular when $p = 2$, $k = m = 1$. For $L_p(u^k)$–Laplacian equation (in (1) $k = m$), see [26])

$$c(m, k, p, k, N, \beta) = \left(|k\gamma_1|^{p-2}k(p-1)\frac{(N-p)\beta - (k(p-1)-1)N}{\beta - (k(p-1)-1)}\right),$$

where $k(p-1) - 1 > 0$.

For p–Laplacian equation ($k = m = 1$)

$$c(1, l, p, 1, N, \beta) = \left(|k\gamma_1|^{p-2}(p-1)\frac{(N-p)\beta - pN}{\beta - p}\right).$$

Notice these results are given in [26] and they are very important for computational aims.

The proofs of Theorems 2 are based on the transformation of Eq. (1) as follows:

$$f(\xi) = \overline{f}(\xi)y(\eta), \quad \eta = \ln\left(a + \xi^{\frac{p}{p-1}}\right)$$

Then with respect to the function $y(\eta)$ we obtain a new nonlinear equation whose solution for $\eta \to \infty$ tends to the constant c indicated in the statement of the theorem.

These results extended to the following equation with variable density

$$\rho_1(x)\frac{\partial u}{\partial t} = div\left(\rho_2(x)u^{m-1}\left|\nabla u^k\right|^{p-2}\nabla u\right) + \rho_1(x)\gamma(t)u^{\beta}, u(0,x) = u_0(x) \geqslant 0, \quad x \in R^N,$$
(12)

where $\rho_1(x) = |x|^{n_1}$, $\rho_2(x) = |x|^{n_2}$, $n_i \in R$, $\nabla(\cdot) - grad(\cdot)$.

Consider the functions defined in Q

$$z_1(t,x) = \bar{u}(t)y(\xi), \quad y(\xi) = (a - \xi^{p/(p-1)})_+^{(p-1)/(k(p-2)+m-1)},$$

$$\xi = \varphi(x)[\tau(t)]^{-1/p}, \quad \varphi(x) = \frac{p-(n_1+n_2)}{p}|x|^{p/(p-(n_1+n_2))}, \quad s = p\frac{N-n_1}{p-(n_1+n_2)}.$$

Theorem 5 *Assume* $k(p-2) + m - 1 > 0$, $n_1 < N$, $n_1 + n_2 < p$,

$$\gamma(t)\tau(t)[\bar{u}(t)]^{\beta-[k(p-2)+m-1]} < s/p, \quad t > 0, \quad u_0(x) \leqslant z_1(0,x), \quad x \in R^N, \quad (13)$$

then the problem (8) is global solvable in Q.

Corollary 2 *Let* $\gamma(t) = 1$, $k(p-2) + m - 1 > 0$. *Then condition of the Fujita type solvability of the problem (10) is*

$$\beta > k(p-2) + m + \frac{N-n_1}{p-(n_1+n_2)}.$$

Corollary 3 *Let* $\gamma(t) = t^{\sigma}$, $k(p-2) + m - 1 > 0$. *Then condition of the Fujita type solvability of the problem (10), (2) is*

$$\beta > \beta_* = (1+\sigma)[k(p-2) + m] + \frac{N-n_1}{p-(n_1+n_2)},$$

value of the critical exponent is equal to

$$\beta = \beta_* = (1+\sigma)[k(p-2) + m] + \frac{N-n_1}{p-(n_1+n_2)}.$$

This result consist all early known results authors [1–6, 18–24] about global solvability problem Cauchy to the degenerate type Eq. (10)

$$c(m,k,p,\sigma,S) = \left(-|k\gamma_1|^{p-2}l(p-1)\frac{(S-p)\beta - (1+\sigma)(k(p-2)+m)S}{\beta - (1+\sigma)(k(p-2))}\right)^{\frac{1}{\beta-(k(p-2)+m)}}$$

$$S = p\frac{N-n_1}{p-(n_1+n_2)} \quad p-(n_1+n_2) > 0, \quad n_1 < N.$$

For the Eq. (10) value of the critical exponent β_* is following

$$\beta = \beta_* = k(p-2) + m + \frac{N - n_1}{p - (n_1 + n_2)}$$

The many works are devoted to the Fujita type critical exponents for semi-linear system of equation

$$\frac{\partial u}{\partial t} = \Delta u + v^{\beta_1}, \quad \frac{\partial u}{\partial t} = \Delta v + u^{\beta_2}$$

Escobedo–Herrero [11] proved the following condition of the critical exponents

$$\frac{\beta_i + 1}{\beta_1 \beta_2 - 1} < N/2, \quad i = 1, 2$$

Below on example cross system we give an algorithm for establishing value of a critical exponent for the system (3) it is find condition of a global solvability using comparison principle, the condition of a finite speed of perturbation which is in particular extension of results of the works for the cross system (3) based on a self-similar approach.

We construct an approximately self-similar system for (3) by following way

$$u(t,x) = \bar{u}(t) w(\tau(t), \eta), \quad v(t,x) = \bar{v}(t) z(\tau(t), \eta), \quad \eta = \int_0^t c(y)dy - x, \quad x \in R^N$$

(14)

$$w(\tau(t), x) = f(\xi), \quad z(\tau(t), x) = \psi(\xi), \quad \xi = |\eta| [\tau(t)]^{-1/p},$$

where $\bar{u}(t) = (T+t)^{-1/(\beta_1 - 1)}$, $\bar{v}(t) = (T+t)^{-1/(\beta_2 - 1)}$, $\tau(t) = \int [\bar{u}(t)]^{(p-2)}$ $[\bar{v}(t)]^{(m_1 - 1)} dt = \int \bar{u}^{(m_2 - 1)} \bar{v}^{(p-2)} dt$,

$$\xi^{1-N} \frac{d}{d\xi} \left(\xi^{N-1} \psi^{m_1 - 1} \left| \frac{df}{d\xi} \right|^{p-2} \frac{df}{d\xi} \right) + \frac{\xi}{p} \frac{df}{d\xi} + b_1 \left(\frac{1}{\beta_1 - 1} f + f^{\beta_1} \right) = 0$$

$$\xi^{1-N} \frac{d}{d\xi} \left(\xi^{N-1} f^{m_2 - 1} \left| \frac{d\psi}{d\xi} \right|^{p-2} \frac{d\psi}{d\xi} \right) + \frac{\xi}{p} \frac{d\psi}{d\xi} + b_2 \left(\frac{1}{\beta_2 - 1} \psi + \psi^{\beta_2} \right) = 0, \quad (15)$$

$$b_1 = (\beta_1 - 1)(\beta_2 - 1)/[(\beta_1 - 1)(\beta_2 - 1) - (m_1 - 1)(\beta_1 - 1) - (\beta_2 - 1)(p-2)],$$
$$b_2 = (\beta_1 - 1)(\beta_2 - 1)/[(\beta_1 - 1)(\beta_2 - 1) - (m_2 - 1)(\beta_2 - 1) - (\beta_1 - 1)(p-2)].$$

If

$$(\beta_1 - 1)(k(p - 2) + m_1 - 1) = (\beta_2 - 1)(k(p - 2) + m_1 - 1)$$

Now consider the functions

From this estimate for the weak solution of the considered problem we have the property of FVPD, i.e.

$$u(t, x) \equiv 0, \ v(t, x) \equiv 0 \ |x| \geqslant l(t) = a[\tau(t)]^{1/p},$$

$$\tau(t) = \frac{(T + t)^{1 - (m_1 - 1)\alpha_2 - (p - 2)\alpha_1}}{1 - (m_1 - 1)\alpha_2 - (p - 2)\alpha_1}, \quad 1 - (m_1 - 1)\alpha_2 - (p - 2)\alpha_1 > 0$$

Since the functions $u_+(t, x)$, $v_+(t, x)$ has the property

$$u_+(t, x) \equiv 0, \quad v_+(t, x) \equiv 0 \ |x| \geqslant l(t) = a^{(p-1)/p}[\tau(t)]^{1/p}.$$

Proof of the Theorem 3 based on comparison principle of solution. For comparison function, we will construct the following Zeldovich–Barenblatt type solution to the main member of the system (3)

$$u_+(t, x) = \bar{u}(t)(T + t)^{-\alpha_1} \bar{f}(\xi), \quad v_+(t, x) = \bar{v}(t)(T + t)^{-\alpha_2} \bar{\psi}(\xi),$$

$$\xi = |x| / [\tau(t)]^{1/p}, \quad \tau(t) = a_1(T + t)^{1/a_1} \ a_1 = 1 - (m_1 - 1)\alpha_2 - k(p - 2)\alpha_1, \quad T > 0,$$

where $\alpha_1 = \frac{n_1}{n_2}\alpha_2$, $n_1 = k(p - 2) - (m_1 - 1)$, $n_2 = k(p - 2) - (m_2 - 1)$, $\alpha_2 = \frac{n_2 N}{pn_2 + [n_1(m_2 - 1)) + k(p - 2)n_2]N}$,

$$\bar{f}(\xi) = A_1 (a - \xi^\gamma)_+^{\gamma_1}, \quad \bar{\psi}(\xi) = A_2 (a - \xi^\gamma)_+^{\gamma_2}, \ A_1 > 0, \quad i = 1, 2, \ a > 0$$

$$\gamma = \frac{p}{p - 1}, \gamma_i = \frac{(p - 1)[k(p - 2) - (m_i - 1)]}{q}, \quad i = 1, 2,$$

$$q = [k(p - 2)]^2 - (m_1 - 1)(m_2 - 1),$$

where constants A_1, A_2 are solution of the system

$$A_1^{p-2} A_2^{m_1-1} = 1/p(k\gamma\gamma_1)^{p-1},$$
$$A_1^{m_2-1} A_2^{p-2} = 1/p(k\gamma\gamma_2)^{p-1}.$$

5 Results of Numerical Experiments and Visualization

At the numerical solution of the considered problems, the equation was approximated on a grid under the implicit circuit of variable directions (for a multidimensional case) in a combination to the method of balance [24]. It is known main problem for numerical solution of considered problem is choice appropriate an initial approximation of solutions depending on value of the numerical parameters. Iterative process were constructed based on the method Picard, Newton and a special method.

Results of computational experiments shows, that all listed iterative methods are effective for the solution of nonlinear problems and leads to the nonlinear effects if we will use as initial approximation the solutions of self-similar equations constructed by the method of nonlinear splitting and by the method of standard equation [21–23]. As it was expected, results of the numerical experiments shows that for achievement of necessary accuracy the method of Newton demands smaller quantity of iterations, than methods of Picard and special method due to a successful choice of an initial approximation. We observe that in each considered cases Newton's method has the best convergence due to good choosing of an initial approximation. The results of all numerical experiments are presented in visual form with animation.

Below are listed typical numerical results the property phenomena a finite velocity of perturbation distribution, and space localization for the solution of the problem (3), (4).

In fast diffusion case $k(p-2) + m_i - 1 < 0$ for computation as an initial approximation were take the function

$$u_0(x,t) = (T+t)^{-\alpha_1}(a+\xi^\gamma)^{\gamma_1}, \quad v_0(x,t) = (T+t)^{-\alpha_2}(a+\xi^\gamma)^{\gamma_2}, \quad k(p-2)+m_i-1 < 0,$$

$$m_1 = 1.1, \quad m_2 = 1.2, \quad p = 1.2, \quad k = 1, \quad n = 0.1, \quad q = 0.2, \quad eps = 10^{-3}.$$

With property $u_0(x,t) = (T+t)^{-\alpha_1}(a+\xi^\gamma)^{\gamma_1}$

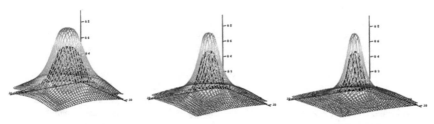

In the slowly diffusion case $k(p-2) + m_i - 1 > 0$ as initial approximation were take the function

$$u_0(x,t) = (T+t)^{-\alpha_1}(a-\xi^\gamma)_+^{\gamma_1}, \quad v_0(x,t) = (T+t)^{-\alpha_2}(a-\xi^\gamma)_+^{\gamma_2}, \quad k(p-2)+m_i-1 > 0,$$

$$m_1 = 5.5 \quad m_2 = 3.2, \quad p = 4.5, \quad k = 2, \quad n = 1.2, \quad q = 1.1, \quad eps = 10^{-3}.$$

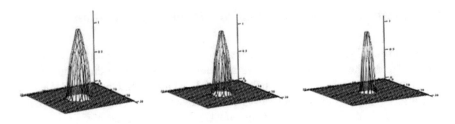

6 Conclusion

Method establishing Fujita type global solvability of the problem Cauchy for one class degenerate type parabolic equation and cross system proved. The algorithm of finding a critical exponent for one degenerate type parabolic equations and cross system is suggested.

The role of a critical exponent for one degenerate type parabolic equations and system is shown.

FVPD properties of solutions diffusion systems with double nonlinearity, with source based on self-similar analysis established

Experimentally showed, that due to base on self-similar analysis of solution some results of numerical experiments keeping nonlinear property of solution as FVPD and space localization.

References

1. Fujita, H.: On the blowing up of solutions to the Cauchy problem for $u_t = \Delta u + u^{1+\alpha}$. J. Fac. Sci. Univ. Tokyo Sect. I(13), 109–124 (1966)
2. Aripov, M.: Asymptotes of the solutions of the non-newton polytrophic filtration equation. ZAMM **80**(3), 767–768 (2000)
3. Martynenko, A.V., Tedeev, A.F.: The Cauchy problem for a quasilinear parabolic equation with a source and inhomogeneous density. Comput. Math. Math. Phys. **47**(2), 238–248 (2007)
4. Martynenko, A.V., Tedeev, A.F.: On the behavior of solutions to the Cauchy problem for a degenerate parabolic equation with inhomogeneous density and a source. Comput. Math. Math. Phys. **48**(7), 1145–1160 (2008)
5. Novruzov, E.: Blow-up phenomena for polytrophic equation with inhomogeneous density and source. J. Math. Phys. **56**, 042701 (2015). https://doi.org/10.1063/1.4916289
6. Martynenko, A.V., Tedeev, A.F., Shramenko, V.N.: The Cauchy problem for a degenerate parabolic equation with an inhomogeneous density and source in the class slowly tending to zero initial functions. Izv. Ross. Akad. Nauk Ser. Math. **76**(3), 139–156 (2012)
7. Zheng, P., Mu, C.: A complete upper estimate on the localization for the degenerate parabolic equation with nonlinear source. Math. Methods Appl. Sci. (2014). Accessed 1 Jan 2014
8. Zheng, P., Mu, C., Ahmed, I.: Cauchy problem for the non-newtonian polytrophic filtration equation with a localized reaction. Appl. Anal. 1–16 (2014). Accessed 20 Feb 2014
9. Zheng, P., Mu, C.: Global existence, large time behavior, and life span for a degenerate parabolic equation with inhomogeneous density and source. Zeitschrift für angewandte Mathematik und Physik **65**(3), 471–486 (2014)

10. Aripov, Í.: Approximate self-similar approach for solving of the quasilinear parabolic equation. Experimentation, Modeling and Computation in Flow Turbulence and Combustion, vol. 2, pp. 9–26. Willey, New York (1997)
11. Escobedo, M., Herrero, M.A.: Boundedness and blow up for a semi linear reaction-diffusion system. J. Differ. Equ. **89**, 176–192 (1991)
12. Marri, Dj.: Nonlinear diffusion equations in biology. Mir, Moscow (1983), 397 p
13. Holodnyok, M., Klich, A., Kubichek, M., Marec, M.: Methods of Analysis of Dynamical Models, p. 365. Mir, Moscow (1991)
14. Kurduomov, S.P., Kurkina, E.S., Telkovskii, : Blow up in two componential media. Math. Model. **5**, 27–39 (1989)
15. Aripov, M., Sadullaeva, Sh: An asymptotic analysis of a self-similar solution for the double nonlinear reaction-diffusion system. Nanosyst. Phys. Chem. Math. **6**(6), 793–802 (2015)
16. Aripov, M., Muhammadiev, J.: Asymptotic behavior of self similarl solutions for one system of quasilinear equations of parabolic type. BuletinStiintific-Universitatea din Pitesti, Seria-Matematica si Informatica **3**, 19–40 (1999)
17. Aripov, M., Rakhmonov, Z.: On the behavior of the solution of a nonlinear multidimensional polytrophic filtration problem with a variable coefficient and nonlocal boundary condition. Contemp. Anal. Appl. Math. **4**(1), 23–32 (2016)
18. Tedeyev, A.F.: Conditions for the existence and nonexistence of a compact support in time solutions of the Cauchy problem for quasilinear degenerate parabolic equations. Sib. Math. J. **45**(1), 189–200 (2004)
19. Vázquez, J.L.: The Porous Medium Equation: Mathematical Theory. Oxford Mathematical Monographs, p. 430. The Clarendon Press, Oxford University Press, Oxford (2007)
20. Aripov, M.: Standard Equation's Methods for Solutions to Nonlinear problems (Monograph), p. 137. FAN, Tashkent (1988)
21. Cho, C.-H.: On the computation of the numerical blow-up time. Jpn. J. Ind. Appl. Math. **30**(2), 331–349 (2013)
22. Zheng, P., Mu, C., Liu, D., Yao, X., Zhou, S.: Blow-up analysis for a quasilinear degenerate parabolic equation with strongly nonlinear source. Abstr. Appl. Anal. **2012**, 19 (2012). https://doi.org/10.1155/2012/109546. Article ID 109546
23. Mersaid, A., Shakhlo, A.S.: To properties of solutions to reaction-diffusion equation with double nonlinearity with distributed parameters. J. Sib. Fed. Univ. Math. Phys. **6**(2), 157–167 (2013)
24. Samarskii, A.A., Galaktionov, V.A., Kurduomov, S.P., Mikhajlov, A.P.: Blowe-up in Quasilinear Parabolic Equation, vol. 4, p. 535. Walter de Grueter, Berlin (1995)
25. Rakhmonov, Z.: On the properties of solutions of multidimensional nonlinear filtration problem with variable density and nonlocal boundary condition in the case of fast diffusion. J. Sib. Fed. Univ. Math. Phys. **9**(2), 236–245 (2016)
26. Aripov, M., Mukimov, A.: An asymptotic solution radially symmetric self-similar solution of nonlinear parabolic equation with source in the second critical exponent case. Acta NUUz **2**(2), 21–30 (2017)

A Language of Terms of Taylor's Formula for Quadratic Dynamical Systems and Its Fractality

Abdulla Azamov

Abstract It is discussing one-step methods of numerical solving of a Cauchy problem for systems of ordinary differential equations. It is constructed an algorithm of numerical solving with arbitrary high precision for quadratic systems based on a context-free grammar of N. Chomsky, that generates a special language of terms of Taylor's formula. The estimation for remainder term is obtained in explicit form. It is studied some combinatorial problems for the language $\mathfrak{I}$ and described its fractal structure. A geometric representation of the fractal is also given in the space l_2.

Keywords Quadratic system · Cauchy problem · Taylor's formula · Chomsky grammar · Fractal · Hilbertian space

1 Introduction

Essential part of results on dynamical systems relies to numerical solution of a Cauchy problem

$$dx/dt = f(x), \qquad x(0) = x_0, \tag{1}$$

where $x \in \mathbb{R}^d$ ([19–21, 27], see also [5]). If a linear case is left aside, quadratic systems with right-side vector-function f given in the form

$$f_i = \sum_{j,k} a_i^{j,k} x_j x_k + \sum_j b_i^j x_j + c_i, \quad i, j, k = 1, 2, \ldots d \tag{2}$$

A. Azamov (✉)
Institute of Mathematics of the Academy of Sciences of Uzbekistan, Tashkent, Uzbekistan
e-mail: abdulla.azamov@gmail.com

© Springer Nature Switzerland AG 2018
A. Azamov et al. (eds.), *Differential Equations and Dynamical Systems*,
Springer Proceedings in Mathematics & Statistics 268,
https://doi.org/10.1007/978-3-030-01476-6_3

make up extraordinary important class in the modern Theory of Dynamical Systems [4, 13, 28]. It contains such famous samples as Lotka–Volterra model [1] and Lorentz and Rössler systems [22, 25–27, 31]. Interest to quadratic systems is also stipulated by the 16th problem of D. Hilbert [24, 30].

For $d \geq 2$ the system (2) is unsolvable in general case. Very rather its properties used being declared basing on numerical solution and computer modeling. (It should be noticed that a picture in a computer's monitor may sensitively depend on quantity of length of a quantization step h. For example if a system has a circle born due to doubled period bifurcation from simple circle then for some values of h there may be observed chaotic trajectory fulfilling Mobius surface with the circle as a boundary while for decreasing h the trajectory may become regular with the circle as its limit set ([6], Chap. 4.).) In any way degree of precision of numerical solution is an important factor for Computational Dynamics.

In most cases variants of Runge–Kutta method are used for numerical solving with order of precision h^s, $s = 2 \div 5$ [6, 8, 32].

Here a new approach to numerical solving of Cauchy problem based on Taylor's formula

$$x(t + h) = \sum_{k=0}^{n} \frac{x^{(k)}(t)}{k!} h^k + R_{n+1}(t, h) \tag{3}$$

will be exposed.

Generally speaking the scheme (3) was not used in practice as an expression for $x^{(n)}$ across the function f and its derivations even more cumbersome. Nevertheless it turns out that the situation becomes essentially simple for quadratic systems. This circumstance allows to give an explicit estimation formula for the remainder term $R_{n+1}(t, h)$ and to construct a simple algorithm for numerical solving with arbitrary high order of precision. The last is based on an interpretation of terms of Taylor's formula as a special language with the context-free grammar of Chomsky [10, 11]. One combinatorial problem of the language will be studied as well. Noteworthy that a tree representing this language has fractal structure. It is suggested a geometrization of this fractal in the Hilbertian space l_2.

2   Estimation of the Remainder Term of Taylor's Formula

Let's begin considerations from the case $d = 1$. Here the equation (1) can be integrated explicitly but arguments will be suitable for high dimensional case. Thus now values of $f(x(t))$, $f'(x(t))$ and $f''(x(t))$ are numbers and in addition $f''(x) = const$. Further these quantities will be called multiplicators and used shortened denotations f, f', f'' for them. Monomials composed by multiplicators and taking part in Taylor's formulae will be called Taylorian terms. Essentially the mentioned names will be used with respect to cases $d \geq 2$ as well when f, f', f'' a not number-valued.

(Keeping in mind just this circumstance the term 'multiplicator' was preferred to 'factor' or 'multiplier'; see below).

Let numbers D_n^k, $n \geq 1$, $k = 0, 1, 2, \ldots, \left[\frac{n-1}{2}\right]$, be defined by recurrent relations

$$D_n^0 = 1, \quad D_{n+1}^k = (n - 2k + 1)D_n^{k-1} + (k + 1)D_n^k \tag{4}$$

($[\cdot]$ denotes the integer part of a number; it is assumed $D_n^k = 0$ for $k < 0$ and $k > [(n-1)/2]$). Note $D_n^1 = 2^{n-1} - n$ so $max_k D_n^k$ increases very quickly when $n \to \infty$.

Statement 1. The following equality holds

$$x^{(n)}(t) = \sum_k D_n^k f''^k f'^{n-2k-1} f^{k+1}. \tag{5}$$

Proof is similar to one for Newton's binomial formula i.e. the method of induction on the parameter n can be provided using (4). It is easier to realize induction step separately for n odd and even as in the first case the number of terms in (5) doesn't change while in the second case it increases to unit. For several beginning values of n the expression (5) looks

$$\dot{x} = f, \quad \ddot{x} = f'f, \quad x^{III} = f''f^2 + f'^2 f, \quad x^{IV} = f'^3 f + 4f''f'f^2,$$

$$x^V = f'^4 f + 11f''f'^2 f^2 + 4f''^2 f^3, \quad \ldots \tag{6}$$

It is useful to note that in each term a power of the multiplicators f, f' will be univalently determined by a value of n and a power of f''.

Now let us consider the case $d \geq 2$. This time values of the multiplicator f is a vector and f' is Jacobi matrix (in other words a tensor $\frac{\partial f_i(x)}{\partial x_j}$ of the rank $(1,1)$) and f'' is a vector with components consisting of bilinear forms (i.e. a tensor $\frac{\partial^2 f_i(x)}{\partial x^j \partial x^k}$ of the rank $(2, 1)$; [9]). Here $f'' = $ const as well but the formula (4) not necessarily true. For example the expression for x^{IV} looks

$$x^{IV} = f'^3 f + 2f''f'ff + f''ff'f + f'f''ff, \tag{7}$$

and if take into account symmetry of the form f'' with respect to contravariant indices i.e. the property $f''(u, v) = f''(v, u)$ then

$$x^{IV} = f'^3 f + 3f''f'ff + f'f''ff. \tag{8}$$

The right side will be written in expanded form as

$$[f'(x)]^3 f(x) + 3f''(x)[f'(x)f(x), f(x)] + f'(x)f''(x)[f(x), f(x)].$$

Now unlike to the case $d = 1$ values of the terms $f''f'ff$ and $f'f''ff$ may differ. For example if $f = (-y^2, x^2)$ then $f''f'ff = -4(x^3y^2, x^2y^3)$, $f'f''ff = $

$-4(y^5, \ x^5)$. That is why Taylorian terms in the expression containing the same power f''^k of the multiplicator f'' won't be assembled to a monomial with the coefficient D_n^k. Nevertheless an essential part of the statement will be saved.

Theorem 1 *The expression for a derivative $x^{(n)}$ of the solution of the Cauchy problem (1) is a sum of collections Δ_n^k of Taylorian terms such that the group Δ_n^k consists of D_n^k monomials from multiplicators f'', f', f in the quantity k, $n - 2k - 1$ and $k + 1$ respectively ($k = 0, 1, \ldots, [(n - 1)/2]$).*

Now suppose that the solution $x(t)$ of the problem (1) exists on a interval $[0, \ T]$ and satisfies there the condition $x(t) \in K$, where T is some given positive number and K is a given compact (and usually convex) subset of the space $\mathbb{R}^d$.

Let

$$M_0 = \max_{x \in K} |f(x)|, \quad M_1 = \max_{x \in K} \|f'(x)\|, \quad M_2 = \|f''(x)\| = \text{const.}$$

(Here all the norms are Euclidean [9]: $\|f'(x)\| = \max_{|u| \leq 1} |f'(x)u|$, $\|f''(x)\| = \max_{|u| \leq 1, \ |v| \leq 1} |f''(x)[u, v]|$). Thus we have

$$|f'u| \leq M_1 |u|, \quad |f''[u, \ v]| \leq M_2 |u| \, |v|.$$

These inequalities imply that all Taylorian terms from the group Δ_n^k admit due to Theorem 1 the same upper estimation by norm in the form $M_2^k M_1^{n-2k-1} M_0^{k+1}$. Hence it is true the following

Theorem 2

$$|R_{n+1}| \leq \frac{h^{n+1}}{(n + 1)!} \sum_k D_{n+1}^k M_0^k M_1^{n-2k} M_2^k, \quad k = 0, 1, \ldots, [(n - 1)/2].$$

Open problem 1. Find expressions for the coefficients D_n^k (similar to the identity for binomial ones $\binom{n}{k} = \frac{n!}{k!(n-k)!}$) and $\sum_k D_n^k$.

3   The Language of Taylorian Terms and an Algorithm for Numerical Solving of a Cauchy Problem with High Precision

As it was noted above if $d \geq 2$ a compact formulae for a derivative $x^{(n)}$ as (5) is difficult to be derived. In order to overcome such an obstacle we apply to Mathematical Linguistics [2]. For that let multiplicators f, f' and f'' redenote by digits 0, 1, 2 respectively. Then each Taylorian term will be inverted to a word over the alphabet

{0, 1, 2}. In order to differ such words from arbitrary words from symbols 0, 1, 2, the first ones will be called d-words. The family of all d-words makes a special language $\mathfrak{I}$. If $\sigma = \varepsilon_1 \varepsilon_2 \ldots \varepsilon_n$ is a d-word, then its length is $|\sigma| = n$ and any part of a form $\varepsilon_m \varepsilon_{m+1} \ldots \varepsilon_l$ $(1 \leq m \leq l \leq n)$ is called a subword.

Usually languages are considered containing an empty word Λ of length 0 that is a subword for any d-word.

The language $\mathfrak{I}$ possesses simple generative grammar ([2], Chap. 4). Indeed the rules $\frac{d}{dt} f(x) = f'(x) f(x)$ (production of a matrix with a vector as a convolution of corresponding tensors of the ranks $(1, 1)$ and $(0, 1)$) and $\frac{d}{dt} f'(x) u = f''(x)[f(x), u]$ (a vector that equals to the value of a vector-valued linear form on vectors $f(x)$ and $u \in \mathbb{R}^d$ that may be considered as a convolution of tensors as well) will be rewritten as generating rules

$$0 \to 10, \qquad 1 \to 20, \qquad\qquad (9)$$

for the language $\mathfrak{I}$. The generation may be begun from the word 0 or from the rule $\Lambda \to 0$. The rules (9) mean that if in a d-word σ 10 and 20 are substituted for pairs 0 and 1 respectively then the result will be a d-word of the length $|\sigma| + 1$. The described generation introduces to $\mathfrak{I}$ a structure of a tree with the beginning

(The last line consists of d-words corresponding to the Taylorian terms from the formula (7)).

Theorem 3 $\mathfrak{I}$ *belongs to the type of context-free languages of N Chomsky.*

Indeed, if 0, 1, 2 are taken as terminal symbols and $\tilde{0}, \tilde{1}, \tilde{2}$ are taken as unterminal ones then context-free grammar [10]

$$\tilde{0} \to \tilde{1}\tilde{0}, \quad \tilde{1} \to \tilde{2}\tilde{0}, \quad \tilde{0} \to 0, \quad \tilde{1} \to 1, \quad \tilde{2} \to 2$$

with initial symbol $\tilde{0}$ generates just d-words and only.

One of the main questions of Mathematical Linguistics concerns a criteria, allowing to define if a given word belongs to the considering language ([2], Sect. 1). For $\mathfrak{I}$ the question has simple answer.

Theorem 4 (on solvability of the $\mathfrak{I}$) *A word σ over the alphabet* {0, 1, 2} *belongs to the language $\mathfrak{I}$ if and only if*

(a) *it ends by the symbol 0;*

(b) *a number of symbols 0 is greater than a number of symbols 2 more on 1;*

(c) *if symbols 2 are numerated in the order from right to left (as in the manner of Arabian writing), then behind jth symbol 2 it should follow exactly $j + 1$ symbols 0.*

Proof The rules (9) imply immediately that every d-word possesses properties (a)–(c). To set sufficiency take a word σ on the alphabet $\{0, 1, 2\}$ with properties (a)–(c). If $|\sigma| = 1$ or 2 then it is obvious that σ is a d-word. Let σ be a word of the length n, $n \geq 3$. Due to property (a) it has a form $\sigma = \rho \varepsilon 0_k$, where 0_k denotes a word consisting of k symbols 0 while $\varepsilon = 1$ or $\varepsilon = 2$ and ρ is some subword of the length $n - k - 1$, $k \geq 1$.

Now if $\varepsilon = 1$ then σ is ended by a subword of the form 10_k (k zeroes following 1). Substituting 0_k for 10_k we get a new word which will still possess properties (a)–(c). Moving back we get σ by the first of rules (9) and so σ belongs $\Im$. Similar reasoning hold for the case $\varepsilon = 2$ as well (now necessarily $k \geq 2$).

The language $\Im$ allows to construct an algorithm (called QS-algorithm below) for solving Cauchy problem of quadratic systems [3]. Let $Extract(n, \sigma; x)$ be a procedure calculating the value of the Taylorian term corresponding to a d-word σ of the length n at a point $x \in K$ (the parameter n is added to the data for convenience only). Thus

$$Extract(1, 0; x) = f(x), \qquad Extract(2, 10; x) = f'(x)f(x),$$

$$Extract(7, 1202010; x) = f'(x)f''(x)\{f(x), f''(x)[f(x), f'(x)f(x)]\}$$

Suppose we are to calculate $x(t + h)$ basing on $x(t)$ with precision h^N. Introduce a transformation D over lists (arrays) from elements of T: if L is any list (array) of d-words then DL is the sequence of lists $D\sigma$, $\sigma \in L$, defining as following: $D\sigma$ is a list of d-words each of them of the length $|\sigma| + 1$ obtained by applying sequentially the rules (9) to all symbols 0 and 1 (in order from left to right). For example

$$D\langle 200 \rangle = \langle 2100, 2010 \rangle, \qquad D\langle 110 \rangle = \langle 2010, 1200, 1110 \rangle$$

and
$$D\langle 200; 110 \rangle = \langle 2100, 2010; 2010, 1200, 1110 \rangle.$$

(In order to highlight lists of d-words, angular parenthesis will be used as above.) Note that the transformation D can be easily realized as procedure in programming languages.

QS-**Algorithm**. Let the order of precision on a step N be given ($N \geq 2$). Take $n = 0$, $L = \langle 0 \rangle$, $H = h$, $S = x(t)$, and put (*) $n = n + 1$. Calculate $\Sigma = \sum_{\sigma \in L} Extract(n, \sigma; x(t))$, $S = S + H\Sigma$. If $n \geq N$, then end process by the value $x(t + h) = S$ else make the new list $L := DL$ and put $H := Hh/n$, then return to the step (*).

One may notice that QS-algorithm itself derives Taylor's formula during calculations due to the language T.

4 A Language of Roots and Combinatorial Analysis of Their Affixes

Let σ be a d-word. Crossing out all symbols 1 in the word σ and the last 0 as well we obtain a word over the alphabet $\{0, 2\}$. It will be called a 2,0-root at that time deleted symbols 1 will be called affixes. We denote $\mathfrak{R}$ the language of all 2,0-roots yielded from d-words.

The notion of a 2,0-root can be linked with Taylor's formula in the following way. Consider some Taylorian term. Since f' is a linear form it should predict a vector-valued quantity. Thus an action of f' (i.e. its convolution with a following tensor of appropriate rank) gives a tensor of the same rank again. Repeating such kind of reductions in the end we get an expression consisting of vectors and bilinear forms only. Its scheme will be expressed by the 2,0-root. This allows to divide the procedure $Extract(n, \sigma; x)$ into two steps: firstly to extract all multiplicators f' by means of convolutions and in parallel find corresponding 2,0-root $\widehat{\sigma}$ and then apply $Extract(2k + 1, \widehat{\sigma}0, x)$ (where k is a number of symbols 2 in the d-word σ). Such a division may make calculation of $Extract(n, \sigma; x)$ faster than the direct process.

The language $\mathfrak{R}$ is solvable as Theorem 2 implies

Statement 2. A word over the alphabet $\{0, 2\}$ will be a 2,0-root iff

(a) ends with the symbol 0;
(b) the number of symbols 0 is equal the number of symbols 2;
(c) if symbols 2 are numerated in the order from right to left, then on the right of the jth symbol 2 follows at least j symbols 0.

Obviously if one inserts the subword 20 into any place (including the beginning and the end) of a 2,0-root he receives a 2,0-root again and conversely any 2,0-root can be obtained in such way beginning from the empty word. The corresponding grammar consists of the following rules:

$$\Lambda \to 20 \quad 0 \to 200, \quad 0 \to 020, \quad 2 \to 200, \quad 2 \to 202 \tag{10}$$

and so $\mathfrak{R}$ has context-free grammar.

One can easily see that the rules (10) may generate concrete 2,0-root in several way (such a property is called ambiguity, [2], Sect. 7). It turns out the language $\mathfrak{R}$ can be built by means of other rules producing all 2,0-roots with unique prehistory and allowing to present $\mathfrak{R}$ as a rooted tree ([16], Sect. 1.5).

Due to Statement 2 every 2,0-root has a form $\rho 0_k$ where a subword ρ ends by the symbol 2. Now define a transformation Π that conforms to $\rho 0_k$ the following list of $k + 1$ words

$$\langle \rho 0_k 20, \quad \rho 0_{k-1} 200, \quad \dots, \quad \rho 200_k \rangle, \tag{11}$$

belonging to $\mathfrak{R}$ because of Statement 2. Π transforms a list of 2,0-roots as well acting to each its element in the same order. Therefore Π can be iterated.

Theorem 5 1°. $\Pi^k(\Lambda)$ *contains only and all 2,0-roots of the length 2k.*
  2°. *All 2,0-roots in the infinite list* $\Pi^1(\Lambda)$, $\Pi^2(\Lambda)$, $\ldots$, $\Pi^k(\Lambda)$, $\ldots$ *are different.*

Proof It is clear that $\Pi^k(\Lambda)$ consists of 2,0-roots of the length $2k$. Let us confirm converse i.e. every 2,0-root σ with the length $2k$ presents in the list $\Pi^k(\Lambda)$. If $|\sigma| = 2$, then $\sigma \in \Pi^1$. Let $k \geq 2$ and σ have a form $\sigma_1 20_m$, $m \geq 1$. Consider a word $\sigma_1 0_{m-1}$ of the length $2(k-1)$, obtained from σ erasing the most right pair of symbols 20. The properties (a)–(c) of the Statement 2 still hold. Thus the supposition $\sigma_1 0_{m-1} \subset \Pi^{k-1}(20)$ implies $\sigma \in \Pi^k(20)$.

The assertion 2° can be also checked easily.

Naturally the following question arises: is it possible to regenerate the list of d-words starting from the list of 2,0-roots? Here partial answer will be given only – we are able to account the collection of d-words of the fixed length with a given 2,0-root but can't restore the whole list of corresponding d-words.

Thus let ρ be a 2,0-root of a length $2k$ and suppose we are to rebuilt a d-word σ such that $|\sigma| = n$. For that $N = n - 2k - 1$ affixes (i.e. symbols 1) should be added. In principle they can be inserted, being distributed someway, in any place in the word ρ i.e. in left side and right side and in the middle between symbols of ρ. A number of such places (will be called boxes) equals $2k + 1$.

Each way of distribution of affixes among boxes will generate a partition of the number N into $2k + 1$ nonnegative addends. In the book [29] such partitions were called compositions. There was also cited that the number of compositions of integer N into K addends equals C_{N+K-1}^{K-1} ([29], Sect. 5.3.1). In our case $N = n - 2k - 1$, $K = 2k + 1$, so that each 2,0-root generates

$$C_{n-2k-1+2k+1-1}^{2k} = C_{n-1}^{2k},$$

many d-words $(k = 0, 1, \ldots, \left[\frac{n-1}{2}\right])$. Note that to complete obtaining d-words 0 should be imputed to the end. For example 2200 generates 15 d-words of the length 7:

$$11\,\square\,2\,\square\,2\,\square\,0\,\square\,0 \rightarrow 112200(0), \ 121200(0), \ 122100(0), \ 122010(0),$$

$$122001(0), \ 2112000, \ 212100(0), \ 212010(0), \ 212001(0), \ 221100(0),$$

$$221010(0), \ 221001(0), \ 220110(0), \ 220101(0), \ 220011(0)$$

(for the first composition boxes are shown, imputed 0 is taken into parenthesis in each composition).

Statement 3. All compositions obtained from 2,0-roots by distribution affixes and considering as words over the alphabet {0, 1, 2} different and belongs to the language $\Im$.

Proof can be provided using induction by a number of imputing affixes.

Note that the algorithm of G.Ehrlich for generation of the list of compositions in lexicographical order is also cited in [29] (Sect. 5.4).

Let α_k denote a number of different 2,0-roots of a length $2k$. Due to Statement 3 there are $\alpha_k \cdot C_{n-1}^{2k}$ different d-words. Remembering that d-words (of course in the form of Taylorian terms) may repeat in the expression for $x^{(n)}$ we get the rough estimation $\alpha_k \cdot C_{n-1}^{2k} \leq D_n^k$.

Open problem 2. Is it possible to find explicit formulae for α_k?

Open problem 3. Find a rule of counting how many times a given d-word σ of the length n is repeated in the expression for $x^{(n)}$ (with or without regard of symmetry of the bilinear form f'').

5   A Fractal of Suffixes

Continuing analysis the language of d-words let us reduce them once more. Every 2,0-root ends with one or more symbols 0 and so it has the form $\rho 0_k$ with a subword ρ ending by the symbol 2. In such situation the subword 0_k will be called a suffix (of the 2,0-root and the corresponding d-word as well). As a result the infinite tree of 2,0-roots turns into a tree of suffixes. The beginning of the reduction looks $20 \rightarrow 0$; $2020 \rightarrow 0$, $2200 \rightarrow 00$; $202020 \rightarrow 0$, $202200 \rightarrow 00$, $220020 \rightarrow 0$, $220200 \rightarrow 00$, $222000 \rightarrow 000$.

The transformation Π defined for generating of 2,0-roots, successfully acts to affixes too and we keep the notation. Thus $\Pi(0_k)$ is the list $\langle 0, 0_2, 0_3, \ldots, 0_{k+1} \rangle$.

It should be noticed some difference between the trees of 2,0-roots and suffixes. In the first case all vertices are different (Theorem 5) but in the second one each suffix 0_k repeats infinitely many times. Moreover we are not able to distinguish them even inside of every list $\Pi^m(0_k)$ with a fixed m as it may also contain the same suffix many times. Therefore the tree-structure is essential for the language of suffixes generated by the transformation Π. This tree will be denoted Γ. In Fig. 1 its beginning is drawn.

It is clear that Γ possesses fractal structure. Indeed first if an iteration number k increases quantity of suffixes 0_m for each k grows rapidly exceeding the geometric progression 2^m. Secondly the whole tree Γ repeats beginning from every vertex, corresponding to the suffix 0 (that follows from acting way of the transformation Π). More commonly let $\widehat{0}_k$ with a fixed k be a concrete suffix met in the list $\Pi^m(0)$ with minimal m. Let $\Gamma(\widehat{0}_k)$ be the subtree growing from the root $\widehat{0}_k$. Then every subtree with a root in a vertex, corresponding to a suffix 0_k with the same k is isomorphic to $\Gamma(\widehat{0}_k)$ (in the sense of Graph theory, [16]).

We get more compact form of the fractal of suffixes if we consider lists $T_k = \langle 0, 0_2, 0_3, \ldots, 0_k \rangle$ instead of suffixes themselves. Here we have a transformation converting each list T_k to the list of lists $\langle T_2, T_3, \ldots, T_{k+1} \rangle$. The obtained tree Δ begins with the list T_1 that will not be met any more in Δ. Figure 2 demonstrates this fractal where every list $\langle T_2, T_3, \ldots, T_k, T_{k+1} \rangle$ is substituted by k i.e. by the quantity of its elements.

The common number of symbols 0 in the list T_k is a triangular number $\frac{k(k+1)}{2}$ therefore the tree in Fig. 3 can be considered as a special fractal of triangular numbers.

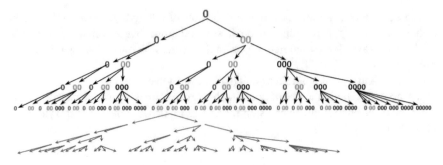

Fig. 1 Fractal of suffixes of Taylorian terms

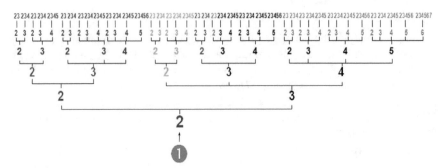

Fig. 2 Arithmatic fractal of suffixes

Fig. 3 Geometrization of
the fractal, $n = 2, \lambda = 0.5$

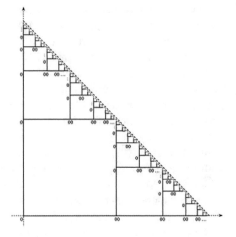

Statement 4. (1) If all subtrees of Γ growing from vertexes 0 are removed (with exception of the root of Γ that should be removed with a unique edge only) then the remainder subgraph will be isomorphic to the fractal Δ. (2) If one moves off all

vertexes 0 with outgoing edges then Γ will be decomposed into a forest of trees each of them isomorphic to Δ.

Above the fractal Γ was represented as an object of Graph Theory and the fractal Δ was done as one of Number Theory. Nowadays number-theoretical fractals becoming usual. Those connected with Pascal's triangle are well studied [7]. Last period searching of fractal properties of prime numbers and zeroes of Riemannian ζ-function and other theoretical fractals becomes more and more intensive [12, 14, 18]. In other hand one of commonwise received types of fractals are those whose self-similarity can be expressed by means of affine contracting transformations [17, 23]. A representation of a fractal using geometric notions can be called 'geometrization'. Here we are going to describe a geometric representation for the fractal Γ.

As a space for that we are to choose infinite dimensional space l_2. (Its points will be distinguished by bold letters.) Let $\mathbf{e}_1, \mathbf{e}_2, \ldots$ be a standard basis so that $\mathbf{e}_i = (\delta_{i1}, \delta_{i2}, \ldots)$ where δ_{ij} is Kronecker's symbol.

Now we fix a value of a parameter λ from the interval $(0, 1)$ and put

$$s_0 = 0; \quad s_m = \lambda + \lambda^2 + \cdots + \lambda^m, \quad m = 1, 2, 3, \ldots; \quad s_\infty = \frac{\lambda}{1-\lambda}.$$

Then we construct an injective map (immersion) $\Phi : \Gamma \to l_2$ by induction on m. First

$\Phi(0) = \mathbf{0} = (0, 0, 0, \ldots)$ for the $\mathbf{0}$ that will play a role of the root of $\Phi(\Gamma)$;

$\Phi(0) = \lambda \mathbf{e}_1 = (\lambda, 0, 0, \ldots)$ and $\Phi(00) = \lambda \mathbf{e}_2 = (0, \lambda, 0, \ldots)$ for the suffixes 0 and 00 respectivly from the list $\Pi(0) = \langle 0, 00 \rangle$;

Φ maps suffixes from $\Pi^2(0) = \langle 0, 00; 0, 00, 000 \rangle$ to the points of l_2

$$(\lambda + \lambda^2, 0, 0, \ldots), \ (\lambda, \lambda^2, 0, \ldots), \ (\lambda^2, \lambda, 0, \ldots), \ (0, \lambda + \lambda^2, 0, \ldots), \ (0, \lambda, \lambda^2, 0, \ldots)$$

respectively.

Suppose that the map Φ is already defined on each suffix $\breve{0}_k \in \Pi^m(0)$, converting it to a point $\mathbf{x}$ that the following conditions are held:

(1) $x_n = 0$ for $n > m + 1$;

(2) $x_n = \sum_{i \in J_n} \lambda^i$ for $n = 1, 2, \ldots, m + 1$ where $J_1, J_2, \ldots, J_{m+1}$ is a partition of the set $\{1, 2, \ldots, m\}$;

(3) $m \in J_k$. (As usual $\sum_{i \in J} \lambda^i = 0$ for $J = \varnothing$).

Now we continue Φ for affixes from the list $\Pi^{m+1}\left(\breve{0}_k\right) = \langle 0, 0_2, \ldots 0_{k+1} \rangle$ taking as their images of corresponding suffixes the points $\mathbf{x} + \lambda^{m+1} \mathbf{e}_1$, $\mathbf{x} + \lambda^{m+1} \mathbf{e}_2$, $\ldots$, $\mathbf{x} + \lambda^{m+1} \mathbf{e}_{k+2}$.

One can easily check that the conditions (1)–(3) holds for newly defined values of Φ.

Let us denote Ξ the image $\Phi(\Gamma) \subset l_2$ with induced structure of a tree.

Theorem 6 *If $0 < \lambda \leq \frac{1}{2}$ then the map Φ is injective: different vertexes of Γ will be mapped to different points.*

Proof is follows of the next inequality: if $J, K \subset \{1, 2, \ldots, m\}$ and

$$\sum_{i \in J} \lambda^i > \sum_{i \in K} \lambda^i,$$

then

$$\sum_{i \in J} \lambda^i > \lambda^{m+1} + \sum_{i \in K} \lambda^i.$$

This is consequence of a simple inequality $\lambda^m > \sum_{i \in K} \lambda^i$ being true for arbitrary finite subset $K \subset \{m+1, m+2, \ldots\}$.

Remark 1 Theorem 6 obviously holds in the case of transcendental λ as well.

In according to construction every point $\mathbf{x} \in \Xi$ (i.e. a vertex of the tree Ξ) has only finite number of coordinates not equal 0 so that the integer

$$\rho(x) = \min\{n \mid x_k = 0 \text{ for all } k > n\}$$

is correctly defined. It will be called *rank* of the point x.

In order to get imagine about construction of the l_2-fractal it is useful to study its finite dimensional sections (briefly "crowns") $\Xi_d = \{x \in \Xi \mid \rho(x) \leq d\}$, $d = 1, 2, \ldots$. Ignoring zero coordinates $x_n = 0$, $n = d+1, d+2, \ldots$, of points from Ξ_d allows us to consider the crowns as finite dimensional objects ("trees") growing in the correspondent space $\mathbb{R}^d$).

The crown Ξ_1 consists of points of the real axis making decreasing geometric progression with the denominator λ. (Ξ_1 may be called "the main trunk" of the l_2-fractal, that stays after cutting all "lateral brunches". Ξ_1 respects to sequence of leftside 0's of the lists $\Pi^m(0)$. It can be considered as the simplest fractal.

The crown Ξ_2 consists of the trunk Ξ_1 and lateral brunches growing from its vertexes in the direction of the axis x_2 so that all its vertexes have a degree 3 besides the origin $(0,0)$ having degree 2. Behavior of the sequence of layers $H^m = \Xi_2 \cap \Phi(\Pi^m(0))$ is more interesting. Its limit in Hausdorff metrics under $m \to \infty$ will be a compact subset $K_2(\lambda)$ of the segment I_1 (one-dimensional simplex) joining points $(\lambda + s_\infty, 0)$ and $(0, \lambda + s_\infty)$. In the case $\lambda = \frac{1}{2}$ the set $K_2(\lambda)$ coincides with I_1 (Fig. 3; here and further x_1-axis is drawn vertically in order to let the tree of l_2-fractal to grow upwards). In the case $\lambda = \frac{1}{3}$ the set $K_2(\lambda)$ is similar to Cantor's fractal, moreover its projections to both coordinate axis coincide with classical Cantor's compact (Fig. 4). If λ less than $\frac{1}{3}$ then the set $K_2(\lambda)$ will be more rare "Cantor's dust" [14, 23].

Fig. 4 Fractal of the crown
$K_3(\lambda)$, $\lambda = 1/3$

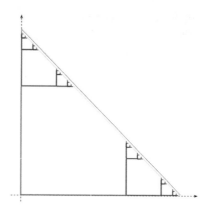

Fig. 5 Fractal of the crown
$K_3(\lambda)$

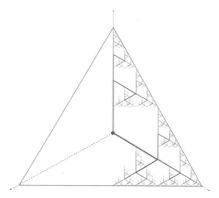

The crown Ξ_3 has more complicated structure but that is typical for higher odd dimensions as well. It has lateral branches growing in the directions of x_1 and x_2 axis only from vertexes of the trunk and of branches which are parallel to the trunk. But now there lateral branches growing in the direction of the axis x_3 from all other vertexes (Fig. 5).

We note that a picture for even-dimensional crowns differs from one for odd dimensional crowns. In the last case projections of coordinate axes can be drawn equiangular (as in Fig. 6 for $d = 5$) while in the first case if $d > 2$ equiangular projections of axes may coincide in pairs. Taking account this circumstance in Fig. 7 the crown Ξ_4 illustrated in nonequiangular projection.

Vertexes belonging to the layer $H_d^m = \Xi_d \cap \Phi\left(\Pi^m(0)\right)$ lay in the simplex $s_m \Delta^{d-1}$ homothetic to the standard one

$$\Delta^{d-1} = \{y \in R^d \mid \sum y_i = 1, \ y_i \geq 0, \ i = 1, 2, \ldots, d\}$$

with the coefficient s_m. At $m \to \infty$ the sequence H_d^m approaches to some compact subset $K_d(\lambda)$ of the limit simplex $s_\infty \Delta^{d-1}$. If $\lambda = \frac{1}{3}$ then $K_d(\lambda)$ will be a subset of

Fig. 6 Fractal of the crown
$K_5(\lambda)$

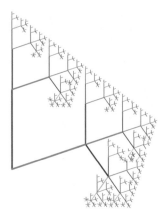

Fig. 7 Fractal of the crown
$K_4(\lambda)$

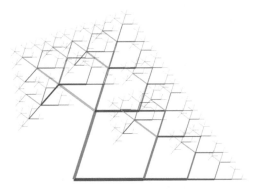

"Sierpinski pyramid" that is a generalization of the two-dimensional case known as "Sierpinski napkin".

Now let us return to the l_2-fractal Ξ. It has a rich semigroup of contracting affine maps $\Pi : \Xi \to \Xi$ such that the image $\Pi(\Xi)$ is similar to Ξ itself. We will constraint ourselves highlighting only one family of self-similarities.

Statement 5. Let A_n be a transformation acting in the space l_2 and compounded from homothety with the coefficient λ^n and shift to the vector $s_n e_1$ ($n = 1, 2, \ldots$). Then A_n maps Ξ into itself such that the image $A_n(\Xi)$ is a subtree with the root in the vertex $s_n e_1$.

Proof Noting that both A_1 and Π are affine maps let us consider their acts on the layer H^m. As A_1 and Π commutes we have "a commutative diagram"

$$
\begin{array}{ccc}
x & \xrightarrow{A_1} & \lambda x + \lambda e_1 \\
\Pi \downarrow & \Pi \downarrow & \\
x + \lambda^m e_j & \xrightarrow{A_1} & \lambda x + \lambda^{m+1} e_j + \lambda e_1
\end{array}
$$

that means A_1 is affine contracting. To end the proof it is sufficient to note that A_n is an iteration of A_1.

Problem 4. Calculate the Hausdorff dimension of the set $K_d(\lambda)$ [15]. Give some characteristics similar to Hausdorff dimension for a fractal $K_\infty(\lambda) = \lim\limits_{d\to\infty} K_d(\lambda)$ that is a compact subset of the Hilbertian simplex $\{\mathbf{x} \in l_2 |\, \|x\| \le 1\}$.

Acknowledgements Author express his gratitude to $\boxed{\text{A.Sh.Kuchkarov}}$ for useful discussions and to O. S. Akhmedov, M. A. Bekimov and A. I. Sotvoldiev for help.

References

1. Ahmad, Sh., Stamova, I.M. (eds.): Lotka-Volterra and Related Systems, 236 p. Harvard University Press (2013)
2. Aho, A.V., Ullman, J.D.: The theory of languages. Math. Syst. Theory **2**(2), 97–125 (1962)
3. Akhmedov, O.S., Bekimov, M.A.: A program realization of Azamov's algorithm for numerical solution of quadratic systems. Official registration certificate for computer programs, No DGU 03767. Republic of Uzbekistan (2016)
4. Artes, J.C., Llibre, J.: Quadratic Hamiltonian vector fields. J. Diff. Equ. **107**, 90–95 (1994)
5. Azamov, A.A., Akhmedov, O.S.: Existence of a complex closed trajectory in a three-dimensional dynamical system. J. Comput. Math. Math. Phys. **51**(8), 1449–1456 (2011)
6. Bakhvalov, N.S.: Numerical Methods, p. 63. Mir Publishers, Moscow (1977)
7. Bondarenko, B.A.: Generalized Pascal Triangles and Pyramids; their Fractals, Graphs, and Applications. The Fibonacci Association, Santa Clara (1993)
8. Butcher, J.C.: Numerical Methods for Ordinary Differential Equations, 2nd edn. Wiley, New York (2008)
9. Cartan, H.: Differential Calculus, p. 160. Kershaw, London (1983)
10. Chomsky, N.: Three models for the description of language. IRE Trans. Inf. Theory **2**, 113–124 (1956)
11. Chomsky, N., Schutzenberger, M.P.: The Algebraic Theory of Context-free Languages. Computer Programmimg and Formal Systems, p. 118. Amsterdam (1969)
12. Chung-Ving, R.: Distribution of the digital units of primes. Chaos, Solitons Fractals **13**, 1215–1302 (2002)
13. Coppel, W.A.: A survey of quadratic systems. J. Diff. Equ. **2**, 293–304 (1966)
14. Crownover, R.M.: Introduction to Fractals and Chaos, p. 306. Jones & Bartlett, Sudbury (1995)
15. Dang, Y., Kauffman, L.H., Sandin, D.: Hypercomplex Iterations, Distance Estimation, and Higher-Dimensional Fractals. World Scientific, Singapore (2002)
16. Diestel, R.: Graph Theory, p. 415. Springer, Berlin (2005)
17. Falconer, K.J.: Techniques of Fractal Geometry, p. 368. Wiley, New York (2014)
18. Folsom, A., Kent, Z.A., Ono, K.: l-adic properties of the partition function. Adv. Math. **229**, 1586–1609 (2012)
19. Guckenheimer, J.: Environments for exploring dynamical systems. Int. J. Bifurc. Chaos **1**, 269–276 (1991)
20. Guckenheimer, J.: Phase portraits of planar vector fields: computer proofs. J. Exp. Math. **4**, 153–165 (1995)
21. Guckenheimer, J.: Numerical Analysis of Dynamical Systems. Handbook of Dynamical Systems. 2, pp. 345–390. Elsevier (2002)
22. Hirsch, M.W., Smale, S., Devaney, R.: Differential Equations, Dynamical Systems, and An Introduction to Chaos, 2nd edn. Academic Press, Boston (2004)

23. Hutchinson, J.E.: Fractals and self similarity. Indiana Univ. Math. J. **30**, 713–747 (1981)
24. Ilyashenko, Yu.: Centennial history of Hilbert's 16th problem. Bull. AMS **39**(3), 301–354
25. Pchelintsev, A.N.: Numerical and physical modeling the dynamics of Lorenz system. Numer. Anal. Appl. **7**, 159–167 (2014)
26. Peitgen, H.O., Jurgens, H., Saupe, D.: Rossler Attractor. Chaos and Fractals: New Frontiers of Science, pp. 636–646. Springer, Berlin (2004)
27. Qiao, Z., Li, X.: Dynamical analysis and numerical simulation of a new Lorenz-type chaotic system. Math. Comput. Model. Dyn. Syst **20**(3), 264–283 (2004)
28. Rein, J.W.: A Bibliography of Qualitative Theory of Quadratic Systems of Differential Equations in the Plane. Delft University of Technology (1987)
29. Reingold, E.M., Nievergelt, J., Deo, N.: Combinatorial Algorithms. Theory and Practice.. Prentice-Hall, Inc. (1977)
30. Smale, S.: Mathematical problems for the next century. Math. Intell. **20**(2), 7–15 (1998)
31. Sparrow, C.: The Lorenz Equation: Bifurcations, Chaos and Strange Attractors, p. 270. Springer, Berlin (1982)
32. Suli, E., Mayers, D.F.: An Intruduction to Numerical Analysis, p. 434. Cambridge University Press, Cambridge (2003)

Discrete-Numerical Tracking Method for Constructing a Poincaré Map

Abdulla Azamov, Akhmedov Odiljon and Tilavov Asliddin

Abstract In this paper the effectiveness and applicability of the DN-tracking method for constructing Poincaré maps are investigated. The DN-tracking method is demonstrated by samples of dynamical systems having closed trajectories, bifurcations of homoclinical loop of a saddle and period doubling.

Keywords Dynamical system · DN-tracking method · Closed trajectory · Poincaré map · Numerical methods · Bifucation

1 Introduction

The task of the qualitative theory of differential equations, as generally accepted, is to study the properties of individual solutions or their families in those cases when they are either can't be integrated in an explicit form or reduced to an equation of lower orders by means of the first integrals. Therefore, according to the approach of A. Poincaré who founded the Theory of Dynamical Systems, the main goal of the Qualitative Theory is to establish certain properties by some methods. To solve such kind of problems various analytical [1–4], as well as topological (geometric) methods have been developed [5].

A. Azamov (✉) · A. Odiljon · T. Asliddin
Institute of Mathematics named after V. I. Romanovskii of the Academy of Sciences
of Uzbekistan, Tashkent, Uzbekistan
e-mail: abdulla.azamov@gmail.com

A. Odiljon
e-mail: odiljon.axmedov@gmail.com

T. Asliddin
e-mail: asliddintm@mail.ru

© Springer Nature Switzerland AG 2018
A. Azamov et al. (eds.), *Differential Equations and Dynamical Systems*,
Springer Proceedings in Mathematics & Statistics 268,
https://doi.org/10.1007/978-3-030-01476-6_4

When the nonlocal properties of a particular solution or the system in general are considered as an object of study, then analytical methods become ineffective. What concerns to topological methods they require the conditions as usual difficult to check. In this regard, as the capabilities of computing technology increase, methods of numerical integration and computer visualization are being more and more employed [6–11]. This approach even has been called "Computational Dynamics" [12], similar to "Topological Dynamics" [13].

In the general case, the use of numerical methods and computer experiments can only play an auxiliary role as heuristic means, at least from the point of view of generally accepted methodology of Mathematics. Namely, each statement, formulated on the basis of an approximate solution or based on the results of a computer experiment, a posteriori should perfectly be proven. This is the main defect of many existing numerical methods. Since, in practice it is very difficult to justify the results of computations rigorously. Ideally, one wants to obtain a rigorously result direct from the computations rather than using them as just heuristic.

The method of discrete-numerical tracking (further briefly DN-tracking), suggested by the first author of this work, is intended for this purpose [14]. The essence of the method is to draw concrete conclusions about the behavior of exact trajectories or their beam on the basis of real and finite amount of information computable by the computer and storable in its memory.

In this article, the possibilities and limitations of the applicability of the DN-tracking method will be discussed and will be surveyed results of research obtained at the department of "Dynamical Systems" of the V.I. Romanovskii Institute of Mathematics of the Academy Sciences of the Republic of Uzbekistan.

2   Discussion

We consider a Cauchy problem

$$\dot{z} = f(z), z(0) = \xi, \tag{1}$$

where $z \in \mathbb{R}^d$ is a polynomial vector-valued function. Further, it will be assumed that $d \geq 2$ and $f(\xi) \neq 0$. The problem (1) has a unique solution $z(t)$ that is called a positive semi-trajectory, defined on some interval $[0, \tau)$, $\tau > 0$.

In the Qualitative Theory of Dynamical Systems arise a number of questions related to the properties of the solution $z(t)$. Let us list some of them for which the DN-tracking method presents its effectiveness.

1°. If an interval $[0, T]$ is given, is it possible to claim that $z(t)$ exists on $[0, T]$ (that is $T < \tau$)?

2°. Suppose that an answer of the question 1 is positive, and a compact subset K of $\mathbb{R}^d$ is given. Does the inclusion $z(t) \in K$ hold on the interval $[0, T]$?

3°. Do the trajectories starting from some neighbourhood of the point ξ return to its neighbourhood? In other words, does the Poincaré map $\Phi : U \to \Gamma$ exist? (Here

Γ is a hyperplane with normal $f'(\xi)$ passing through the point ξ and U is some neighbourhood of ξ in Γ.)

4°. Suppose that an answer of the question 3 is also positive. Is there a closed trajectory passing near the point ξ? (More generally, does the map Φ have periodic points?).

5°. Let an answer of the question 4 be positive as well, i.e. the system has a closed trajectory. Will it be a limit cycle? More generally, what kind of properties does the map Φ possess?

Of course, this list can be continued. For example, questions about the existence of a homoclinic loop, a heteroclinic cycle, an invariant torus, various kinds of bifurcations may take place. All the above issues are related to the global properties of dynamical systems for which analytical methods are not always effective.

This circumference causes broad application of numerical methods and computer visualization. For example, one of the coryphaeuses of the theory of dynamical systems D.V. Anosov wrote [15] that an enormous amount of research devoted to the Lorentz system [16] might be divided into two groups:

- Researchers of the first group have been assuming a priori the existence of a Poincaré map and based on such supposition derived one or another property of the system (see, for example, [17]).
- Researchers of the second group conclude about the property of the system using numerical solution, mainly from the results of computer experiments (see, for example, [18]).

Obviously, every statement formulated on the basis of a numerical solution must be strictly proven [19]. In some cases, such a justification can be obtained, ignoring the fact that they have been formulated by means of approximate methods. In the method of DN-tracking both formulations and proofs of such statements are carried out based on a numerical solution combining with deductive argument. Efficiency and justification of the method essentially depend on the estimates between exact and numerical solutions. This circumstance limits the possibilities of approximate methods and so DN-tracking method. It is well-known that the most explicit estimates exist for one-step methods, so further we will deal only with them, mainly with different schemes of the Runge–Kutta method [10].

In one-step methods the problem (1) is replaced by the recurrent scheme

$$z_{n+1} = z_n + h F (h, z_n), z_0 = \xi \tag{2}$$

where h is the parameter of discretization, F is the operator expressing the scheme of the concrete form of the method, for example, $F(h, z) = z + \frac{1}{6} (k_1 + 4k_2 + k_3), k_1 = hf(z), k_2 = hf(z + 0.5k_1), k_3 = hf(z + 2k_2 - k_1)$, for the scheme of Runge–Kutta method of the 3rd order.

For a scheme of form (2) one has an estimation [10]

$$|z(nh) - z_n| \le Ce^{LT} h^s = \varepsilon_1, \tag{3}$$

where z_n is defined by (2), $L = \max\limits_{z \in K} \|\partial f / \partial z\|$ is the Lipschitz constant, $h = T/N$ and s is the order of accuracy of the method (usually $s = 2 \div 5$), C is a constant depending on f, F, s, K and T. Further the sequence z_n will be called *a discrete trajectory*.

It should be specially emphasized that estimations of type (3) are correct only in the case of positive answers to questions 1° and 2°, which are formulated above.

If system (1) is nonlinear, then it is also impossible to calculate the sequence z_n precisely. For example, even in the case $z_{n+1} = z_n + hz_n^2$, $z_0 = 1$, $h = 0.01$ one is not able to find exact value of z_{1000}. Therefore, we deal with the final sequence $\{\zeta_n\}$, $n = 0, 1, 2, \ldots, N$, obtained by approximate calculations by means of a computer and the results of the calculations are stored in its memory with a certain precision.

Now let us discuss the relation between z_n and ζ_n. It is well-known [10, 11]

$$|z_n - \zeta_n| \leq \frac{e^{LT} - 1}{Lh} \Delta = \varepsilon_2, \tag{4}$$

where Δ is the rounding error in one step and depends on the function f and the configuration of a used computer. It is very important that the estimation (4) not only forbids $h \to 0$, but prevents the excessive decrease of h [20].

The estimates (3) and (4) imply an inequality

$$\varepsilon_1 \varepsilon_2^s \geq \tilde{C}(f, T, \Delta), \tag{5}$$

that can be interpreted as a kind of uncertainty principle in the computational dynamics.

The categorical conclusion follows from inequality (5): even if the answers of the questions 1° and 2° are positive, there may be no guaranteed connection between the exact trajectory $z(t)$ and the sequence ζ_n.

3 The Paradigma of the DN-Tracking Method

There are may be such a combination of circumstances allowing to choose a method F and an interval of $[0, T]$ and a length of step h and a positive number ε such that

(1) it is possible to establish the existence of a solution $z(t)$ on the interval $[0, T]$;
(2) to prove that the relations $z(t) \in K$, $z_n \in K$, $\zeta_n \in K$ hold for $t \in [0, T]$ and $n \in \{0, 1, \ldots, N\}$, $n = [t/h]$ for given T and K;
(3) to prove the existence of a Poincaré map;
(4) to derive some of its properties, for example, the existence of periodic points.
    The paradigma of the method is expressed in the following way.
    Let $K_0 = \{z \in \mathbb{R}^2 | \ |z_i - a_i| \leq \alpha_i, i = 1, 2\}$ be a fixed parallelepiped. K_1, K_2, K_3 are other parallelepipeds such that $K_j \subset K_{j+1}$ and dist $(K_j, \partial K_{j+1}) = \frac{\varepsilon}{3}$, $j = 0, 1, 2$, $\varepsilon > 0$.

Assumption A. $\zeta_n \in K_0$ for $n = 0, 1, 2, \ldots, N$ (since K_0 is a parallelepiped, this inclusion can be checked by a computer).

Theorem 1 *Suppose that the following estimations are established.*
1. $|\zeta_n - z_n| < \frac{\varepsilon}{3}$, *as long as* $z_n \in K_1$;
2. $|z(nh) - z_n| < \frac{\varepsilon}{3}$, *as long as* $z(nh) \in K_2$;
3. $|z(t) - z(nh)| < \frac{\varepsilon}{3}$, *as long as* $z(t) \in K_3$, $(n = [t/h])$.
Then assumption A implies that all inequalities are true for all $n = 1, 2, \ldots, N$ *and* $t \in [0, T]$.

The theorem easily can be proved "by contradiction method".

Corollary 1 $z(t) \in \operatorname{Int} K_3$ *and* $|z(t) - \zeta_n| < \varepsilon$ *for all* $t \in [0, T]$
Under Assumption A in two-dimensional systems it is enough to track one positive semi-trajectory in order to construct a "Bendixson's bag" [2] instead of Poincaré map.

In order to construct a Poincaré map in multidimensional systems in contrast to two-dimensional systems, it is necessary to track a definite beam of trajectories.

If $U \subset \Gamma$, (Γ was defined above), we take a uniform rectangular grid $M_\delta = \delta I_\delta$, where $I_\delta = \{(i_1, i_2, \ldots, i_d) \in \mathbb{Z}^d | (\delta i_1, \delta i_2, \ldots, \delta i_d) \in \overline{U}\}$. Set $\zeta_v^{(n)}$, $z_v^{(n)}$ and $z_v(t)$ denote the numerical, discrete and exact trajectories starting from a point $v \in M_\delta$, respectively. Further the trajectory starting from a point $u \in \overline{U}$ is denoted by $z_u(t)$. It is clear that for an arbitrary point u one can choose v such that $|u - v| < \delta$.

Consider a net of parallelepipeds K_j, $j = \overline{0, 4}$ such that $K_{j+1} = K_j \overset{*}{+} \frac{\varepsilon}{4} D$, where $K_0 = \{z \in \mathbb{R}^d | |z_i - a_i| \leq \alpha_i, i = 1, 2 \ldots d\}$ is a fixed parallelepiped, D is the unit parallelepiped in $\mathbb{R}^d$, $\overset{*}{+}$ is Minkowskii sum [21]. Note that $\operatorname{dist}(K_j, \partial K_{j+1}) = \frac{\varepsilon}{4}$, $K_j \subset K_{j+1}$, $j = 0, 1, 2, 3$.
Assumption B. $\zeta_v^{(n)} \in K_0$ for all $n = 0, 1, 2, \ldots, N$ and $v \in M_\delta$.

In every concrete case this assumption can be checked by a computer executing limited number of arithmetic operations and comparisons.

Theorem 2 *Suppose that*
1. $|z_v^{(n)} - \zeta_v^{(n)}| < \frac{\varepsilon}{4}$ *as long as* $z_v^{(n)} \in K_1$;
2. $|z_v(nh) - z_v^{(n)}| < \frac{\varepsilon}{4}$ *as long as* $z_v(nh) \in K_2$;
3. $|z_v(t) - z_v(nh)| < \frac{\varepsilon}{4}$ *as long as* $z_v(t) \in K_3$, $(n = [t/h])$;
4. $|z_u(t) - z_v(t)| < \frac{\varepsilon}{4}$ *as long as* $z_u(t) \in K_4$.
Then assumption B implies that all these inequalities are true for all $n = 1, 2, \ldots, N$, $t \in [0, T]$.
Particularly

$$z_u(t) \in \operatorname{Int} K_4$$

and $z_u(t)$ *exists on the interval* $[0, T]$. *Moreover*

$$|z_u(t) - \zeta_v^{(n)}| < \varepsilon$$

for all $t \in [0, T]$.

Thus, by means of a finite sequence of numerical d-vectors $\zeta_0, \zeta_1, \ldots, \zeta_N$ stored in the memory of a particular computer, one can trace from an exact trajectory $z(t), 0 \leq t \leq T$ with an accuracy of ε. If the relations and the conditions of the theorem are satisfied, then there is a chance to establish the existence of a Poincaré map.

Below some examples of application of the DN-tracking method will be given. In all of them, a special technique is used additionally, allowing significantly to increase the range of applicability of DN-tracking.

As one can see from the inequalities (3), (4) the efficiency of the numerical solution depends essentially on a value of the factor e^{LT}, which can bring to naught the smallness of the quantities h^s and Δ. For example, if we consider Lorentz system with parameters $\sigma = 10$, $\beta = 8/3$, $\rho = 28$ and take only $T = 1.1$ and parallelepiped $K = \{(x, y, z) \mid -9.5 \leq x \leq 20; -10 \leq y \leq 27.7; 0 \leq z \leq 48.4\}$ that a priori containing the attractor, then it turns out $e^{LT} \approx 5.8 \cdot 10^{15}$ [22]. But $T = 1.1$ is too small to draw a conclusion about chaos and a strange attractor in the Lorentz system.

This circumstance has been met when the DN-tracking method was used for the first time. It was observed that if one divides the interval $[0, T]$ into several parts if that the estimation (3) becomes more effective [23]. This effect can be explained relying on the fact that if one divides $[0, T]$ into two parts, the factor e^{LT} decreases with the rate of a geometric progression, while the computational errors for reduced segments add up only. In all the examples being considered below, this technique showed its usefulness for constructing a Poincaré map.

3.1 Existence of a Closed Trajectory

Consider a dynamical system simulating the chemical reaction of I. Prigogine, known as "Brusselator" [24]

$$\begin{aligned} \dot{x}_1 &= a + x_1^2 x_2 - (b+1)x_1 \\ \dot{x}_2 &= bx_1 - x_1^2 x_2 \end{aligned} \tag{6}$$

where a and b are positive parameters.

Shifting the origin of Cartesian system Ox_1x_2 to the fixed point $P = (a, b/a)$ by the formula $z_1 = x_1 - a, z_2 = x_2 - b/a$ we get

$$\begin{aligned} \dot{z}_1 &= (b-1)z_1 + a^2 z_2 + \frac{b}{a}z_1^2 + 2az_1^2 z_2 + z_1^2 z_2 \\ \dot{z}_2 &= bz_1 - a^2 z_2 - \frac{b}{a}z_1^2 - 2z_1^2 z_2 - z_1^2 z_2. \end{aligned} \tag{7}$$

The computer experiment provides a basis for the heuristic assertion that the system (7) with $a = 1$ and $b = 2.01$ has a closed trajectory of the period $T \approx 6.28$ in the domain $K = \{(z_1, z_2) \mid -0.33 \leq z_1 \leq 0.53; -0.62 \leq z_2 \leq 0.53\}$. In this example $M_0 < 4.08$, $11.52 < M_1 = L < 11.54$, $M_2 < 14.37$ and $e^{LT} \approx 10^{31}$.

Fig. 1 Consruction of the
Poincaré map for (6)

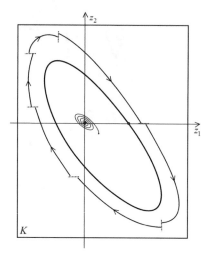

The division of the interval $[0, T]$ into 6 equal parts yields the estimates $|z_n - \zeta_n| \leq 1.5 \cdot 10^{-7}$, $|z(nh) - \zeta_n| \leq 4.2 \cdot 10^{-5}$, $|z(t) - z(nh)| \leq 4.2 \cdot 10^{-6}$, allowing us to construct the Poincaré map $\Phi : (0, 0.32] \to (0, 0.32]$ as a composition of six monodromy mappings (see Fig.1; for details [25]).

In this case, the DN-tracking method guarantees that the Poincaré map has the property: $\Phi(0.32) < 0.32$. Thus, a positive semitrajectory started from the point 0.32 forms a "Bendixson's bag". Since for $a = 1$ and $b = 2.01$ the point $(0, 0)$ is the only singular point inside the "bag" that is an unstable focus, Poincaré-Bendixson theorem implies the following

Theorem 3 *For $a = 1$ and $b = 2.01$, the system (7) in the domain*

$$K = \{(z_1, z_2)| -0.33 \leq z_1 \leq 0.53; -0.62 \leq z_2 \leq 0.53\}$$

has a closed trajectory with period T, $6.27 < T < 6.29$.

Note that the DN-tracking method does not allow one to prove the uniqueness of a closed trajectory. This can be proven, for example, by the L.A. Cherkass method [26].

3.2 Existence of the Bifurcation of Homoclinic Loop

Consider the following two-dimensional nonlinear system with one nonlinear term [27]

$$\begin{aligned} \dot{x} &= ax + y + x^2, \\ \dot{y} &= bx + y, \end{aligned} \tag{8}$$

that is can be considered as a model for bifurcations.

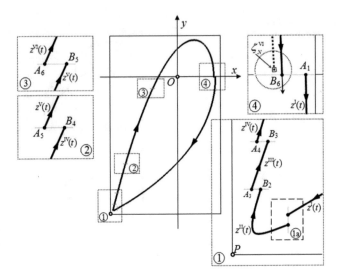

Fig. 2 A cycle of (6) close to the homoclinic loop

Fig. 3 Homoclinic loop of a
saddle in (6)

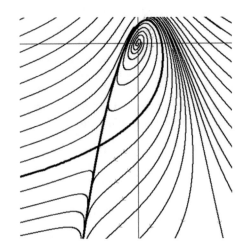

The DN-tracking method allows to construct a "Bendixon's bag" from $a = -0.72$
till $a = -0.74$, when the cycle period increases to 16.72 (Fig. 2). Computer calculations lead us to the heuristic conclusion that if a increases further then the closed
trajectory gets destroyed and as a result the bifurcation of homoclinical loop of
separatrices of a saddle occurs (Fig. 3).

   It is easy to verify that when the parameter a passes from the domain $a < -1$ to the
domain $a > -1$ for $b \in (-\infty, -1)$, the Poincaré–Andronov–Hopf bifurcation takes
place in the system (8). Thus, for a greater than -1, but close to it, system (8) has
a closed trajectory. The Poincaré–Andronov–Hopf bifurcation technique guarantees

Fig. 4 Bifurcation diagram for (6)

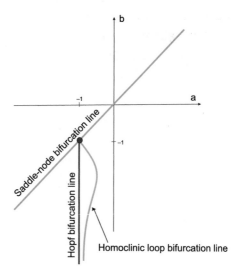

the existence of a such bifurcation for a narrow range of values a near the line $a = -1$ only.

DN-tracking method allows to prove the following non-local result.

Theorem 4 *Set* $b = -2.005$. *Then there exists values* $\tilde{a} \in (-0.74, -0.72)$ *and* $\bar{a}, \bar{a} > \tilde{a}$ *for that the following properties are hold:*

(a) if $-1 < a < \tilde{a}$ *system (8) has a closed trajectory;*
(b) if $\tilde{a} < a < \bar{a}$ *then (8) does not have a closed trajectory.*

Hence, on the basis of the fact that the DN-tracking method has a "assurance factor" in constructing the Poincaré map (in this example "Bendixon's bag"), it is obvious that the bifurcation of homoclinic loop takes place for values close to $b = -2.005$. Covering the interval $-3.5 < b < -1$ with a sufficiently fine grid of such intervals one may construct a bifurcation curve of the homoclinic loop of a saddle for system (8).

On the whole, it can be shown there are three types of bifurcations of codimension 1 in system (8), namely Poincaré–Andronov–Hopf, saddle-nodes and homoclinic loops bifurcations (Fig. 4) [28].

3.3 Existence of a Closed Trajectory of Multidimensional Systems

The DN-tracking method allows to proof existence of a closed trajectory for multidimensional systems as well. As noted above, unlike two-dimensional systems of the

Fig. 5 Consruction of the
Poincaré map for (9)

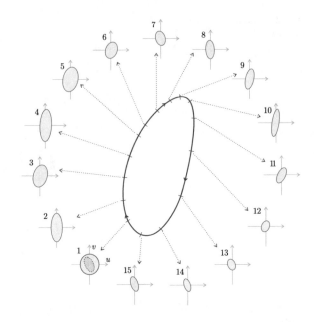

type (7), (8), in multidimensional systems it is necessary to track some beam of trajectories. In [29] this was realized for the three-dimensional model of the brusselator [30], given by the system

$$\begin{aligned}
\dot{z}_1 &= z_1^2 z_2 - z_1(z_3 + 1) + 1 \\
\dot{z}_2 &= z_1 z_3 - z_1^2 z_2 \\
\dot{z}_3 &= -z_1 z_3 + \alpha
\end{aligned} \tag{9}$$

Taking $K = \{(z_1, z_2, z_3) \,|\, 0.7 \le z_1 \le 1.5; 0.7 \le z_2 \le 1.9; 0.9 \le z_3 \le 1.7\}$, $\alpha = 1.25$, $\xi = (0.7679, 1.3730, 1.4226)$, $T = 8.6$ and dividing $[0, T]$ into 15 parts, it is possible to establish the existence of a Poincaré map $\Phi : S \to \Gamma$ such that $\Phi(S) \subset S$ and dist$[\partial S, \Phi(S)] > 10^{-5}$, where S is a disk with radius $5 \cdot 10^{-4}$ in the plane Γ (Fig. 5).

A more interesting example is given by the system [31]

$$\begin{aligned}
\dot{x}_1 &= -x_2 - x_3 + x_1^2 - x_2^2 - x_3^2 \\
\dot{x}_2 &= x_1 - x_3 - x_1^2 \\
\dot{x}_3 &= x_2.
\end{aligned} \tag{10}$$

DN-tracking method following computer calculations allows establishing that in the region

$$K = \{(x_1, x_2, x_3) \,|\, -0.7 \le x_1 \le 0.4; -0.8 \le x_2 \le 0.8; -0.9 \le x_3 \le 0.3\}$$

Fig. 6 Consruction of the
Poincaré map for (10)

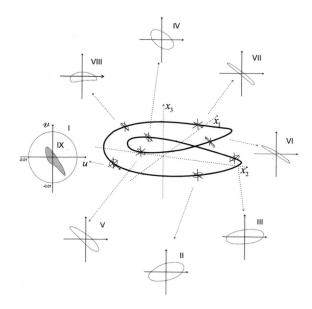

system (10) has a closed trajectory of period $T \approx 8.18$.

Here the segment $[0, T]$ divides into 8 parts. The scheme of construction of the
Poincaré map $\Phi : S \rightarrow \Gamma$ is demonstrated in the Fig. 6.

3.4 Existence of the Period Doubling Bifurcation

The appearance of a closed trajectory of the the system (10) remembering a boundary
of Mobius sheet make us to suggest that it should be a result of a period doubling bifurcation from a simple closed trajectory. It is obvious that if we perturb the coefficient

Fig. 7 Simple closed
trajectory

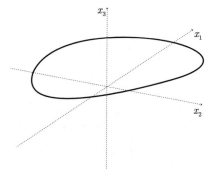

Fig. 8 Period doubling bifurcations for (11)

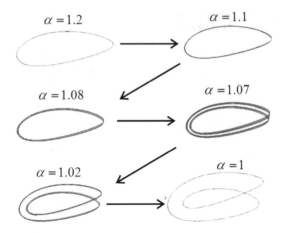

at the term x_3 in the second equation, in other words, take it as $\dot{x}_2 = x_1 - 1.25x_3 - x_1^2$ then there exists a simple closed trajectory of the period ≈ 3.9 (Fig. 7).

Therefore, homotopy joining the last and the initial systems

$$\begin{aligned}
\dot{x}_1 &= -x_2 - x_3 + x_1^2 - x_2^2 - x_3^2 \\
\dot{x}_2 &= x_1 - \alpha x_3 - x_1^2 \\
\dot{x}_3 &= x_2
\end{aligned} \tag{11}$$

should contain demanded bifurcation.

Theorem 5 *In system (11) the period doubling bifurcation takes place for some* $\alpha \in (1, 1.2)$ *(Fig. 8.)*

References

1. Poincaré, H.: Sur les courbes définies par les équations différentielles (III). Journal de mathématiques pures et appliquées 4e série, tome **1**, 167–244 (1885)
2. Andronov, A.A., Leontovich, E.A., Gordon, I.I., Maier, A.G.: Qualitative Theory of Second-Order Dynamic Systems. Nauka, Moscow (1966). Wiley, New York (1973)
3. Arnold, V.I., Ilyashenko, Y.S.: Ordinary differential equations. In: Advances in Science and Technology: Modern Problems in Mathematics: Fundamental Directions,vol. 1, pp. 7–149. VINITI, Moscow (1985). (in Russian)
4. Shilnikov, L.P., Shilnikov, A.L., Turaev, D.V., Chua, L.O.: Methods of Qualitative Theory in Nonlinear Dynamics, Part I, p. 419. Ijevsk, Moscow (2004). (in Russian)
5. Arnold, V.I.: Geometrical Methods in the Theory of Ordinary Differential Equations. Springer, Berlin (1988)
6. Griffiths, D.F., Higham, D.J.: Numerical Methods for Ordinary Differential Equations. Springer Undergraduate Mathematics Series (2010)
7. Holmes, M.H.: Introduction to Numerical Methods in Differential Equations, vol. 52, p. 239. Springer, New York (2007)

8. Boldo, S., Faissole, F., Chapoutot, A.: Round-off Error Analysis of Explicit One-Step Numerical Integration Methods. https://hal.archives-ouvertes.fr/hal-01581794. Submitted on 5 Sep 2017
9. Kehlet, B., Logg, A.: A posteriori error analysis of round-off errors in the numerical solution of ordinary differential equation. The article is published with the open access at Springerlink.com (2016)
10. Bakhvalov, N.S.: Numerical Methods. M: Nauka (1973). (in Russian)
11. Butcher, J.C.: Numerical Methods for Ordinary Differential Equations, p. 463. Wiley, Chichester (2008)
12. Guckenheimer, J.: Computational Tools for Dynamical Systems: An Overview (2008)
13. Sibirrskii, K.S.: Introduction to Topological Dynamics. Monograph, p. 143 (1970). (in Russian)
14. Azamov. A.A.: DN-tracking method for proving the existence of limit cycles. In: Abstracts of Papers of the International Conference on Differential Equations and Topology dedicated to L.S. Pontryagin on the occasion of his 100th birthday, pp. 87–88. Moscow (2008). (in Russian)
15. Anosov, D.V.: Encyclopedia of mathematics. M.: Soviet encyclopedia I.M.Vinogradov (1977–1985). (in Russian)
16. Lorenz, E.N.: Deterministic nonperiodic flow. J. Atmos. Sci. **20**(2), 130–141 (1963)
17. Klinshpont, N.E.: On the topological classification of Lorenz-type attractors. Sbornik: Math. **107**(4), 547–594 (2006)
18. Ghys, É.: Knots and dynamics. In: Proceedings of the International Congress of Mathematicians, Madrid, , vol. 1, pp. 247–277. 22–30 August (2006)
19. Tucker, Warwick: A rigorous ODE Solver and Smale's 14th problem. Found. Comput. Math. **2**, 53–117 (2002)
20. Heisenberg, W.: Über den anschaulichen Inhalt der quantentheoretischen Kinematik und Mechanik. Zeitschrift fúr Physik (in German) **43**(3–4), 172–198 (1927)
21. Nikolskii, M.S.: The first direct method of L.S. Pontryagin in differential games. MGU, Moscow (1984). (in Russian)
22. Kehlet, B., Logg, A.: Quantifying the Computability of the Lonenz System. arXiv:1306.2782v1 [math.NA] 12 Jun 2013
23. Azamov, A.A., Tilavov, A.M.: A simplest nonlinear system having limit cycle. Uzbek Math. J. No. 2, pp. 35–41 (2009)
24. Glandsdorff, P., Prigogine, I.: Thermodynamic Theory of Structure, Stability and Fluctuations. Wiley, New York (1971)
25. Azamov, A.A., Ibragimov, G., Akhmedov, O.S., Ismail, F.: On the proof of existence of a limit cycle for the Prigogine brusselator model. J. Math. Res. **3**(4), 983–989 (2011)
26. Cherkas, L.A.: The method of support functions in the problem of estimating the number of limit cycles of autonomous systems in the plane. In: Abstracts of Papers of the International Conference on Differential Equations and Topology dedicated to L.S. Pontryagin on the occasion of his 100th birthday, pp. 207–208. Moscow, (2008). (in Russian)
27. Tilavov, A.M.: On a limit cycle of model nonlinear system near saddle homoclinical loop bifurcation. Uzbek mat. J. No. 3, pp. 131–137 (2010)
28. Tilavov, A.M.: On a simple quadratic dynamical system with three classical types of bifurcations. In: Doklady of the Academy of Science of Uzbekistan, no. 5, pp. 3–5 (2011)
29. Azamov, A.A., Akhmedov, O.S.: Existence of a closed trjectory in a three dimensional model of brusselator. In: Applied Mathematics and Mechanics (2018) [in print]
30. Hairer E., Nõrsett S., Wanner G.: Solving Ordinary Differential Equations I, Nonstiff Problems, p. 528. Springer, Berlin (2003)
31. Azamov, A.A., Akhmedov, O.S.: Existence of a complex closed trajectory in a three-dimensional Dynamical System. Comput. Math. Math. Phys. **51**(8), 1353–1359 (2011)

A Discrete Mathematical Model for Heat Transfer Process in Rotating Regenerative Air Preheater

Bekimov Mansur and Fathalla A. Rihan

Abstract In this paper we propose a discrete mathematical model for heat transfer process in two-layer rotating regenerative air preheaters of a thermal power plant. The model is formulated by discretizing the process as a result of averaging both temporal and spatial variables. We take into account partial mixing of gas and air. Some conditions that ensure the asymptotic stability of the discrete system have also been deduced.

Keywords Regenerative air preheater · Heat transfer · Discrete system Mathematical model · Stability

1 Introduction

An air preheater (APH) is a general term to describe any device designed to heat air before another process (for example, combustion in a boiler) with the primary objective of increasing the thermal efficiency of the process. They may be used alone or to replace a recuperative heat system or to replace a steam coil. The purpose of the air preheater is to recover the heat from the boiler flue gas which increases the thermal efficiency of the boiler by reducing the useful heat lost in the flue gas. As a consequence, the flue gases are also sent to the flue gas stack (or chimney) at a lower temperature, allowing simplified design of the ducting and the flue gas stack. It also

B. Mansur (✉)
Institute of Mathematics of the Academy of Uzbekistan, 100170 Mirzo Ulugbek street, Tashkent, Uzbekistan
e-mail: mansu@mail.ru

F. A. Rihan
Department of Mathematical Sciences, College of Science, United Arab Emirates University, 15551 Al-Ain, United Arab Emirates
e-mail: frihan@uaeu.ac.ae

© Springer Nature Switzerland AG 2018
A. Azamov et al. (eds.), *Differential Equations and Dynamical Systems*,
Springer Proceedings in Mathematics & Statistics 268,
https://doi.org/10.1007/978-3-030-01476-6_5

allows control over the temperature of gases leaving the stack (to meet emissions regulations, for example).

There are two types of regenerative air preheaters: the rotating-plate regenerative air preheaters (RRAP) and the stationary-plate regenerative air preheaters. The rotating-plate design (RRAP) consists of a central rotating-plate element installed within a casing that is divided into two (bi-sector type), three (tri-sector type) or four (quad-sector type) sectors containing seals around the element. The seals allow the element to rotate through all the sectors, but keep gas leakage between sectors to a minimum while providing separate gas air and flue gas paths through each sector. It is usually attached to a thermal power plant (TPP) in order to increase the efficiency of the station by means of preheating of air blasting into stations boiler. Heating method is carried out due to the heat of combustion products of the fuel, i.e. a hot mixture of smoke and gas (hereinafter simply called gas). Also, using of RRAP allows us to reduce impact of the thermal pollution of the atmosphere [1].

In this work, we consider the case of a unit with the main block of cylindrical shape which consists of a rotor rammed by corrugated thin steel and flat sheets on the surface where the heat exchange occurs. The rotor is divided into sections, separated by plane sheets slowly rotates in a stationary casing. The rotor space is divided into two halves, hot and cold, throughout an imaginary plane passing through the axe of the cylinder. The gas is fed across the hot half in one direction while air flows across the cold half in the opposite direction. During continuous rotation of the rotor, its metal running alternately passes through the gas and air flows and as a result, the heat of the gases is then transferred into the air.

Observation and control of both temperature of the plates, and air and gas emerging from RRAP are effective operations of RRAP [1–3]. Direct measurement of the temperature of incoming and outgoing air and gas is easily carried out; while monitoring and controlling the temperature of the plates requires a sophisticated measuring technique. It is worth mentioning that the temperature regime of the plates plays an important role in the prevention of corrosion. That is why, in practice, mathematical modeling of the heat transfer process in RRAP is essential and widely utilized.

Over the last decades, various mathematical models have been proposed to model heat transfer process in the RRAP [1–6]. Generally, mathematical equations of thermal conductivity and mass transfer can be used to model the operations of RRAP [2]. However, several challenges and difficulties arise which affect the efficiency of the model. One challenge is the configuration of the plates which have a complicated geometry, that practically makes it impossible to formulate the boundary conditions. The second difficulty is that the air and gas flows passing through the RRAP rotor are considered turbulent [3]. The need to record the rotation of the rotor is an additional complication. Due to these and other features, all mathematical models of RRAP are constructed under some simplifying assumptions and ignoring some realities.

Some new approaches have been proposed in [7, 8] to model the thermodynamic process in RRAP. These models are based on the discretization of both the rotor volume and its rotation, followed by averaging over the space and time variables characterizing the heat exchange process between the plates on one side and air and gas on the other side. In [7], the authors used a simple discrete model, and demonstrated the possibility of finding the values of the parameters characterizing the entire heat exchange process, based on the results of measuring temperature of incoming and outgoing gas and air. In [8], more adequate multi-sectional rotor model has been investigated and studied, particularly, the solvability of the inverse problem was established.

We should mention here that in the real conditions of RRAP operation, there is a small amount of mixing of gas with air and air with gas at the interface between the gas and air parts of the rotor. However, single-layer models, considered in [7, 8], did not take into account this feature. Herein, in this work, we propose a two-layer mathematical model of the heat transfer process in RRAP, which allows this useful feature.

2 Mathematical Model

The formulation of the model is based on discretization process of both temporal and spatial variables. We assume that the inner space of the rotor is divided into two layers (lower layer and upper layer) by a plane perpendicular to the drum axis. Further, we divide the same space conditionally into $4m$ equal sectors by flat surfaces passing through the drum axis, so that each layer contains $2m$ sectors. We define the average temperature metal ramming of the lower layer as x_i, and of the upper layer as y_i, $i = 1, 2, \ldots, 2m$. The values $i = 1, 2, \ldots, m$ relate to sectors located on the "cold" half of the rotor through which the air flow passes and the values $i = m + 1, m + 2, \ldots, 2m$—relate to the "hot" half through which the gas flow passes in the opposite direction. We also denote the average temperature of the of the i-th section of the lower layer as u_i, and the upper layer as—v_i, $i = 1, 2, \ldots, 2m$.

The process of heat exchange between air and gas flows through the metal ramming of RRAP is extremely complicated, which is difficult to properly formulate it as an initial boundary value problem of a system of differential equations. In order to overcome this difficulty/complexity, we discretize the problem toward time [8]. With this aim, we select a discretizing subinterval h ($h > 0$), so that at time h drum turns to angle $\frac{\pi}{m}$. At this time, the sections will be swapped in a cyclic order.

We assume the fowling process of heat exchange in the RRAP: At time $t = nh$, $n = 0, 1, 2, \ldots$, the values of the parameters, introduced above, equal $x_i(n), y_i(n), u_i(n), v_i(n)$. We assume that the drum, and gas and air flows remain stationary during time interval $[nh, (n + 1)h)$, and heat exchange takes place in each section, which connects the values of $x_i(n)$ and $u_i(n)$, and of $y_i(n)$ and $v_i(n)$, $i = 1, 2, \ldots, 2m$. Hereafter, at time $t = (n + 1)h$ the drum rotates abruptly to the angle $\frac{\pi}{m}$, in the section of the lower layer with the numbers $i = 1, 2, \ldots, m$ air enters, displacing the previously filled heat carrier with the temperature $u_i(n)$ to the corresponding upper layer, and in the upper section with the numbers $i = m + 1, m + 2, \ldots, 2m$ the gas flows in, displacing the portion of the heat carrier in the bottom layer. At the same time, the heat carrier occupying the sections of the upper layer with numbers $i = 1, 2, \ldots, m$ ($i = m + 1, m + 2, \ldots, 2m$ i.e.,

warmed up air with a gas admixture) is fed into the boiler of the TPP, and the coolant
occupying the sections of the lower layer with numbers (i.e., a cooling gas with a
small fraction of air) flows out to atmosphere through the furnace tube.

Let us discuss the process of mixing gas and air in the proposed model: The gas
fill the section of the lower layer with the number $2m$, in accordance with the above
assumptions goes to the section with the number of $i = 1$ the lower layer, then it is
forced into the top layer section with the same number, then in the next step, mixing
with heated air in sections of the upper layer with numbers $i = 2, 3, \ldots, m$, blown
in boiler. Similarly, the portion of air that occupies the section with the numbers
$i = m$ of the upper layer feeds into the $(m + 1)$ section of the lower layer and in the
next step flows into the tube, mixing with the gas of the sections with the numbers
$i = m + 2, \ldots, 2m$ of the lower layer.

To make it easy, we consider the following assumptions:

Assumption 1 The heat exchange at each interval $[nh, (n + 1)h)$ occurs in accordance with Newton's linear law.

Assumption 2 For sections of number $m + 1$ of the lower layer, and of number 1
of the upper layer, where air and gas are mixed, heat exchange between the nozzles
and local heat carriers does not take place.

Under these assumptions, we arrive at the following equations describing our
model:

$$
\begin{aligned}
x_1(n + 1) &= \bar{\beta} x_{2m}(n) + \tilde{\beta} u_{2m}(n), \\
x_i(n + 1) &= \bar{\alpha} x_{i-1}(n) + \tilde{\alpha} p(n), \quad i = 2, \ldots, m + 2, \\
x_{m+2}(n + 1) &= \bar{\alpha} x_{m+1}(n) + \tilde{\alpha} u_{m+1}(n), \\
x_i(n + 1) &= \bar{\beta} x_{i-1}(n) + \tilde{\beta} u_{i-1}(n), \quad i = n + 3, \ldots, 2m, \\
y_1(n + 1) &= \bar{\beta} y_{2m}(n) + \tilde{\beta} q(n), \\
y_2(n + 1) &= \bar{\beta} y_1(n) + \tilde{\beta} v_1(n), \\
y_i(n + 1) &= \bar{\alpha} y_{i-1}(n) + \tilde{\alpha} v_{i-1}(n), \quad i = 3, \ldots, m + 1, \\
y_i(n + 1) &= \bar{\beta} y_{i-1}(n) + \tilde{\beta} q(n), \quad i = m + 2, \ldots, 2m.
\end{aligned}
\tag{1}
$$

for the temperature of metal part of the sectors and

$$
\begin{aligned}
u_{m+1}(n + 1) &= \tilde{\gamma} y_m(n) + \bar{\gamma} v_m(n), \\
u_i(n + 1) &= \bar{\delta} y_{i-1}(n) + \tilde{\delta} q(n), \quad i = m + 2, \ldots, 2m, \\
v_1(n + 1) &= \bar{\delta} x_{2m}(n) + \tilde{\delta} u_{2m}(n), \\
v_i(n + 1) &= \tilde{\gamma} x_{i-1}(n) + \bar{\gamma} p(n), \quad i = 2, \ldots, m + 1
\end{aligned}
\tag{2}
$$

for the temperature of the heat carriers. Here $\tilde{\alpha} = \alpha h$, $\bar{\alpha} = 1 - \tilde{\alpha}$, $\tilde{\beta} = \beta h$, $\bar{\beta} = 1 - \tilde{\beta}$, $\tilde{\gamma} = \gamma h$, $\bar{\gamma} = 1 - \tilde{\gamma} h$, $\tilde{\delta} = \delta h$, $\bar{\delta} = 1 - \tilde{\delta}$ and $\alpha, \beta, \gamma, \delta$ are the parameters
characterizing the heat exchange process in RRAP (geometry and heat capacity of
the ramming and rotor casing, composition, density and humidity of air and gas, heat
capacity and thermal conductivity coefficients e.c.), $p(n)$ is the average temperature
of the air entering the RRAP, and $q(n)$ is the average temperature of the incoming
gas.

In the meantime, the average temperature of the air entering the boiler of the TPP from the RRAP is given by the formula

$$V(n) = \frac{1}{m} \sum_{i=1}^{m} v_i(n),$$

and the average temperature of the gas leaving for the atmosphere is provided by

$$U(n) = \frac{1}{m} \sum_{i=m+1}^{2m} u_i(n).$$

In the systems (1)–(2) the parameters α, β, γ, δ carry characteristics of heat exchange on the contact surfaces of nozzles with heat carriers. It must be noted that their value can vary during the process, for example, due to nozzle wear, soot deposits, deviation of heat transfer from linear law, and do on. They can be influenced by heat exchange on the outer case of RRAP. Here it well be assumed that the values α, β, γ, δ remain unchanged. Under this assumption, relations (1) represent a closed system of linear discrete equations [9]. It must be specially emphasized that the derived system does not belong to the type of difference equations obtained from differential equations as a result of replacing the derivatives by the incremental ratio, since for $h \to 0$ equality (1)–(2) don't go over to a system of differential equations. This circumstance is associated with the rotation of the RRAP rotor and therefore is considered a mathematical expression of this feature [8].

Equations (1) and (2) represent a discrete linear system of inhomogeneous difference equations, of order $6m$ referred to unknown

$$x_1(n), \ldots, x_{2m}(n), y_1(n), \ldots, y_{2m}(n), u_{m+1}(n) \ldots, u_{2m}(n), v_1(n), \ldots, v_m(n).$$

We can put this system in a matrix form

$$z(n+1) = Az(n) + r(n), \tag{3}$$

with

$$r(n) = \left(\tilde{\alpha} p(n)I, \ \tilde{\alpha} p(n)J, \ \tilde{\beta} q(n)J, \ \tilde{\beta} q(n)I, \ \bar{\delta} q(n)I, \ \bar{\gamma} p(n)I \right)^{T},$$

where $I = (0, 1, \ldots, 1)^T$, $J = (1, 0, \ldots, 0)^T$ (T is a transposition sign, turning a row-vector into a column-vector).

We arrive at the following Theorem.

Theorem 1 *If* $h < \min\left(\frac{1}{\alpha}, \frac{1}{\beta}, \frac{2\alpha}{\alpha^2+\gamma^2}, \frac{2\beta}{\beta^2+\delta^2} \right)$ *then the system (3) is asymptotically stable.*

Proof The assertion is equivalent to the fact that all eigenvalues of the matrix A lay inside the unit circle, which in its turn is equivalent to the estimation $\|A\| < 1$ (for Euclidean norm of the matrix). We denote $(Ax)_k$ the component of number k of the vector Ax. Then, using Cauchy inequality we have

$$
\begin{aligned}
(Ax)_1^2 &= \left(\bar{\beta} x_{2m} + \tilde{\beta} u_{2m}\right)^2 \le \left(\bar{\beta}^2 + \tilde{\beta}^2\right)\left(x_{2m}^2 + u_{2m}^2\right), \\
(Ax)_{k+1}^2 &= \bar{\alpha}^2 x_k^2, \ k = 1, \ldots, m, \\
(Ax)_{m+2}^2 &= (\bar{\alpha} x_{m+1} + \tilde{\alpha} u_{m+1})^2 \le \left(\bar{\alpha}^2 + \tilde{\alpha}^2\right)\left(x_{m+1}^2 + u_{m+1}^2\right), \\
(Ax)_{k+1}^2 &= \left(\bar{\beta} x_k + \tilde{\beta} u_k\right)^2 \le \left(\bar{\beta}^2 + \tilde{\beta}^2\right)\left(x_k^2 + u_k^2\right), \ k = m+2, \ldots, 2m-1, \\
(Ax)_{2m+1}^2 &= \bar{\beta}^2 y_{2m}^2, \\
(Ax)_{2m+2}^2 &= \left(\bar{\beta} y_1 + \tilde{\beta} v_1\right)^2 \le \left(\bar{\beta}^2 + \tilde{\beta}^2\right)\left(y_1^2 + v_1^2\right), \\
(Ax)_{2m+k+1}^2 &= \left(\bar{\alpha}^2 y_k + \tilde{\alpha} v_k\right)^2 \le \left(\bar{\alpha}^2 + \tilde{\alpha}^2\right)\left(y_k^2 + v_k^2\right), \ k = 2, \ldots, m, \\
(Ax)_{2m+k+1}^2 &= \bar{\beta}^2 y_k^2, \ k = m+1, \ldots, 2m-1, \\
(Ax)_{3m+k+1}^2 &= \bar{\delta}^2 y_k^2, \ k = m+1, \ldots, 2m-1, \\
(Ax)_{5m+k+1}^2 &= \bar{\gamma}^2 x_k^2, \ k = 1, \ldots, m-1.
\end{aligned}
\tag{4}
$$

If we set $C = \max\left\{\bar{\alpha}^2 + \tilde{\alpha}^2, \ \bar{\beta}^2 + \tilde{\beta}^2, \ \bar{\alpha}^2 + \tilde{\gamma}^2, \ \bar{\beta}^2 + \tilde{\delta}^2\right\}$ the system of inequalities (4) yields

$$
|Ax|^2 \le C \, |x|^2 .
$$

Under the condition of the theorem $\alpha h < 1$, therefore

$$
\bar{\alpha}^2 + \tilde{\alpha}^2 = (1 - \alpha h)^2 + (\alpha h)^2 = 1 - 2\alpha h(1 - \alpha h) < 1.
$$

Similarly $\bar{\beta}^2 + \tilde{\beta}^2 < 1$. Further, $(\alpha^2 + \gamma^2)h < 2\alpha$, so

$$
\bar{\alpha}^2 + \tilde{\gamma}^2 = (1 - \alpha h)^2 + (\gamma h)^2 = 1 - h[2\alpha - (\alpha^2 + \gamma^2)h] < 1.
$$

Also $\bar{\beta}^2 + \tilde{\delta}^2 < 1$. Consequently $C < 1$. The proof is complete.

As a consequence of the above Theorem, we also arrive at the following facts:

Corollary 1 *If $r(n)$ is a bounded sequence, then each solution of equation (4) is also bounded.*

Corollary 2 $E - A$ *is reversible.*

Corollary 3 *Let $\lim_{n \to \infty} r(n) = l$. Then each solution $z(n)$ approaches the limit $(E - A)^{-1}l$ as $n \to \infty$ independently of $z(0)$.*

For the proof, we refer to similar ones given in [7].

Acknowledgements This work is funded by Ministry of Innovation Development of the Republic of Uzbekistan (research project OT-F4-84). The authors are most grateful to Professor A. Azamov for his valuable comments and numerous suggestions concerning the content of the paper.

References

1. Kirsanov, Yu.A.: Cyclic Thermal Processes and the Theory of Thermal Conductivity in Regenerative Air Heaters. Fizmatlit, Moscow (2007). (in Russian)
2. Kovalevskii, V.P.: Simulation of heat and aerodynamic processes in regenerators of continuous and periodic operation. I. Nonlinear mathematical model and numerical algorithm. J. Eng. Phys. Thermophys. **77**(6), 1110–1118 (2004)
3. Chi-Liang, L.: Regenerative air preheaters with four channels in a power plant system. J. Chin. Inst. Eng. **32**(5), 703–710 (2009)
4. Alagic, S., Kovacevic, A., Buljubasic, I.: A numerical analysis of heat transfer and fluid flow in rotary regenerative air pre-heaters. Strojnivski Vestnik **51**(7–8), 411–417 (2005)
5. Heidari-Kaydan, E.H.: Three-dimensional simulation of rotary air preheater in steam power plant. Appl. Therm. Eng. **73**, 397–405 (2014)
6. Drobnic, B., Oman, J., Tuma, M.: A numerical model for the analysis of heat transfer and leakages in a rotary air preheater. Int. J. Heat Mass Transf. **49**, 5001–5009 (2006)
7. Azamov, A.A., Bekimov, M.A.: Simplified model of the heat exchange process in rotary regenerative air pre-heater. Ural Math. J. **2**, 27–36 (2017)
8. Azamov, A.A., Bekimov, M.A.: Discrete model of the heat exchange process in rotating regenerative air preheaters. Prooc. Steklov Inst. Math. **23**, 12–19 (2017)
9. Lakshmikantham, L., Trigiante, D.: Theory of Difference Equations: Numerical Methods and Applications. Basel, New York (1988)

On Attractors of Isospectral Compressions of Networks

Leonid Bunimovich and Longmei Shu

Abstract In the recently developed theory of isospectral transformations of networks isospectral compressions are performed with respect to some chosen characteristics (attributes) of the network's nodes (edges). Each isospectral compression (when a certain characteristic is fixed) defines a dynamical system on the space of all networks. It is shown that any orbit of such dynamical system which starts at any finite network (as the initial point of this orbit) converges to an attractor. This attractor is a smaller network where the chosen characteristic has the same value for all nodes (or edges). We demonstrate that isospectral compressions of one and the same network defined by different characteristics of nodes (or edges) may converge to the same as well as to different attractors. It is also shown that a collection of networks may be spectrally equivalent with respect to some network characteristic but nonequivalent with respect to another. These results suggest a new constructive approach which allows us to analyze and compare the topologies of different networks.

Keywords Isospectral transformations · Spectral equivalence · Attractors

1 Introduction

Arguably the major scientific buzzword of our time is a "Big Data". When talking about Big Data people usually refer to (huge) natural networks in communications, bioinformatics, social sciences, etc, etc, etc. In all cases the first idea and hope is to somehow reduce these enormously large networks to some smaller objects while keeping, as much as possible, information about the original huge network.

In practice almost all the information about real-world networks is contained in their adjacency matrices [1, 2]. An adjacency matrix of a network with N elements

L. Bunimovich · L. Shu (✉)
School of Mathematics, Georgia Institute of Technology, Atlanta, GA 30332-0160, USA
e-mail: lshu6@math.gatech.edu

L. Bunimovich
e-mail: bunimovh@math.gatech.edu

© Springer Nature Switzerland AG 2018
A. Azamov et al. (eds.), *Differential Equations and Dynamical Systems*,
Springer Proceedings in Mathematics & Statistics 268,
https://doi.org/10.1007/978-3-030-01476-6_6

is the $N \times N$ matrix with zero or one elements. The (i, j) element equals one if there is direct interaction between the elements number i and number j of a network. In the graph representation of a network this corresponds to the existence of an edge (arrow) connecting node i to node j. Otherwise an (i, j) element of the adjacency matrix of a network equals zero. It is very rare [1, 2] that the strength of interaction of the element (node) i with the element (node) j is also known. In such cases a network is represented by a weighted adjacency matrix where the (i, j) entry corresponds to the strength of this interaction instead of to 1.

Therefore the problem of compressing a network is essentially a problem of compressing its weighted adjacency matrix. It is a basic fact of linear algebra that all the information about a matrix is contained in its spectrum (collection of all eigenvalues of a matrix) and in its eigenvectors and generalized eigenvectors.

Recently a constructive rigorous mathematical theory was developed which allows us to compress (reduce) matrices and networks while keeping ALL the information regarding their spectrum and eigenvalues. This approach was successfully applied to various theoretical and applied problems [3]. The corresponding transformations of networks were called Isospectral Transformations. This approach is not only limited to the compression of networks. It also allows one to grow (enlarge) networks while keeping stability of their evolution (dynamics), etc (see [3, 4]).

In the present paper we further develop this approach by demonstrating that isospectral compressions generate a dynamical system on the space of all networks. We prove that such a dynamical system converges to an attractor which is a smaller network than the network which was an initial point (network) of this orbit. To create this dynamical system we need to first select some characteristic of the network's nodes (or edges). Then we pick a subset of nodes (edges) based on this characteristic. We then reduce the network onto the subset we just picked. We repeat this procedure and get a dynamical system. It is important to mention that the current graph theory is lacking classification of all graphs which have the same characteristic of the all nodes even for such basic and simplest characteristics as inner and outer degrees. Clearly any complete graph where any two nodes are connected by an edge (in case of undirected graphs) or by two opposite edges (in case of directed graphs) has the same value of any characteristic at any node. Therefore all complete graphs are attractors of any isospectral contraction. However, there are other attractors as well for any characteristic and there is no general classification or description of these attractors. However one can find such attractors when dealing with a concrete network. Therefore, this procedure is a natural tool for analysis of real-world networks. We demonstrate that by choosing different characteristics of either nodes or edges of a network one typically gets different attractors. The structure of such networks gives us new important information about a given network.

We also discuss the notions of weak and strong spectral equivalences of networks and show that classes of equivalence with respect to a weak spectral equivalence consists of a countable number of classes of strongly spectrally equivalent networks. Our results could be readily applicable to analysis of any (directed or undirected, weighted or unweighted) networks.

2 Isospectral Graph Reductions and Spectral Equivalence

In this section we recall definitions of the isospectral transformations of graphs and networks.

Let $\mathbb{W}$ be the set of rational functions of the form $w(\lambda) = p(\lambda)/q(\lambda)$, where $p(\lambda), q(\lambda) \in \mathbb{C}[\lambda]$ are polynomials having no common linear factors, i.e., no common roots, and where $q(\lambda)$ is not identically zero. $\mathbb{W}$ is a field under addition and multiplication [3].

Let $\mathbb{G}$ be the class of all weighted directed graphs with edge weights in $\mathbb{W}$. More precisely, a graph $G \in \mathbb{G}$ is an ordered triple $G = (V, E, w)$ where $V = \{1, 2, \ldots, n\}$ is the *vertex set*, $E \subset V \times V$ is the set of *directed edges*, and $w : E \to \mathbb{W}$ is the *weight function*. Denote by $M_G = (w(i, j))_{i,j \in V}$ the *weighted adjacency matrix* of G, with the convention that $w(i, j) = 0$ whenever $(i, j) \notin E$. We will alternatively refer to graphs as networks because weighted adjacency matrices define all static (i.e. non evolving) real-world networks. Also we will be using "vertex" and "node" interchangeably.

Observe that the entries of M_G are rational functions. Let's write $M_G(\lambda)$ instead of M_G here to emphasize the role of λ as a variable. For $M_G(\lambda) \in \mathbb{W}^{n \times n}$, we define the spectrum, or multiset of eigenvalues to be

$$\sigma(M_G(\lambda)) = \{\lambda \in \mathbb{C} : \det(M_G(\lambda) - \lambda I) = 0\}.$$

Notice that we count the multiplicities of the eigenvalues, i.e. the set $\sigma(M_G(\lambda))$ can have more than n elements, some of which can be equal to each other.

A path $\gamma = (i_0, \ldots, i_p)$ in the graph $G = (V, E, w)$ is an ordered sequence of distinct vertices $i_0, \ldots, i_p \in V$ such that $(i_l, i_{l+1}) \in E$ for $0 \leq l \leq p - 1$. The vertices $i_1, \ldots, i_{p-1} \in V$ of γ are called *interior vertices*. If $i_0 = i_p$ then γ is a *cycle*. A cycle is called a *loop* if $p = 1$ and $i_0 = i_1$. The length of a path $\gamma = (i_0, \ldots, i_p)$ is the integer p. Note that there are no paths of length 0 and that every edge $(i, j) \in E$ is a path of length 1.

If $S \subset V$ is a subset of all the vertices, we will write $\overline{S} = V \setminus S$ and denote by $|S|$ the cardinality of the set S.

Definition 1 (*structural set*) Let $G = (V, E, w) \in \mathbb{G}$. A nonempty vertex set $S \subset V$ is a structural set of G if

- each cycle of G, that is not a loop, contains a vertex in S;
- $w(i, i) \neq \lambda$ for each $i \in \overline{S}$.

In particular, if a structural set S also satisfies $w(i, i) \neq \lambda_0, \forall i \in \overline{S}$ for some $\lambda_0 \in \mathbb{C}$, then S is called a λ_0-structural set.

Definition 2 Given a structural set S, a *branch* of (G, S) is a path $\beta = (i_0, i_1, \ldots, i_{p-1}, i_p)$ such that $i_0, i_p \in V$ and all $i_1, \ldots, i_{p-1} \in \overline{S}$.

We denote by $\mathcal{B} = \mathcal{B}_{G,S}$ the set of all branches of (G, S). Given vertices $i, j \in V$, we denote by $\mathcal{B}_{i,j}$ the set of all branches in $\mathcal{B}$ that start in i and end in j. For each branch $\beta = (i_0, i_1, \ldots, i_{p-1}, i_p)$ we define the *weight* of β as follows:

$$w(\beta, \lambda) := w(i_0, i_1) \prod_{l=1}^{p-1} \frac{w(i_l, i_{l+1})}{\lambda - w(i_l, i_l)}. \tag{1}$$

Given $i, j \in V$ set

$$R_{i,j}(G, S, \lambda) := \sum_{\beta \in \mathcal{B}_{i,j}} w(\beta, \lambda). \tag{2}$$

Definition 3 (*Isospectral Reduction(Compression)*) Given $G \in \mathbb{G}$ and a structural set S, the reduced adjacency matrix $R_S(G, \lambda)$ is the $|S| \times |S|$-matrix with the entries $R_{i,j}(G, S, \lambda), i, j \in S$. This adjacency matrix $R_S(G, \lambda)$ on S defines the reduced graph which is the isospectral reduction of the original graph G.

Remark 1 We will use the terms "reduction" and "compression" interchangeably. One can check that for a graph with complex number weights, the complement of any single node is a structural set. For any subset A of nodes of this network G, it is always possible to isospectrally compress the network G to a network whose nodes belong to A by removing the nodes in the complement of A one after another.

Now we recall the notion of spectral equivalence of networks (graphs).

Let $\mathbb{W}_\pi \subset \mathbb{W}$ be the set of rational functions $p(\lambda)/q(\lambda)$ such that $\deg(p) \leq \deg(q)$, where $\deg(p)$ is the degree of the polynomial $p(\lambda)$. And let $\mathbb{G}_\pi \subset \mathbb{G}$ be the set of graphs $G = (V, E, w)$ such that $w : E \to \mathbb{W}_\pi$. Every graph in $\mathbb{G}_\pi$ can be isospectrally reduced over any nonempty subset of its vertex set [3].

Two weighted directed graphs $G_1 = (V_1, E_1, w_1)$ and $G_2 = (V_2, E_2, w_2)$ are *isomorphic* if there is a bijection $b : V_1 \to V_2$ such that there is an edge e_{ij} in G_1 from v_i to v_j if and only if there is an edge $\tilde{e}_{ij}$ between $b(v_i)$ and $b(v_j)$ in G_2 with $w_2(\tilde{e}_{ij}) = w_1(e_{ij})$. If the map b exists, it is called an *isomorphism*, and we write $G_1 \simeq G_2$.

An isomorphism is essentially a relabeling of the vertices of a graph. Therefore, if two graphs are isomorphic, then their spectra are identical. The relation of being isomorphic is reflexive, symmetric, and transitive; in other words, it's an equivalence relation.

The notion of spectral equivalence of graphs was introduced in [3]. This is the idea that two networks G and H are spectrally equivalent if they reduce to isomorphic graphs in one step, over subsets of vertices selected by a rule τ (e.g. nodes whose inner degrees are less than 2). Then in [5] a less restrictive notion of generalized spectral equivalence of graphs (networks) was introduced. Namely, two networks are weakly spectrally equivalent if they reduce to isomorphic graphs in a finite number of steps (not necessarily the same number of steps) under the same rule for subset selection.

A proof of the following theorem can be found in [5].

Theorem 1 (Generalized Spectral Equivalence of Graphs) *Suppose that for each graph $G = (V, E, w)$ in $\mathbb{G}_\pi$, τ is a rule that selects a unique nonempty subset*

$\tau(G) \subset V$. *Let R_τ be the isospectral reduction of G onto $\tau(G)$. Then R_τ induces an equivalence relation $\sim$ on the set $\mathbb{G}_\pi$, where $G \sim H$ if $R_\tau^m(G) \simeq R_\tau^k(H)$ for some $m, k \in \mathbb{N}$.*

Remark 2 Observe that we do not require $\tau(G)$ to be a structural subset of G. However there is a unique isospectral reduction [3] (possibly via a sequence of isospectral reductions to structural sets if $\tau(G)$ is not a structural subset of G) of G onto $\tau(G)$.

The notion of generalized spectral equivalence of networks (graphs) is weaker than the one considered in [3], where it was required that $m = k = 1$. Therefore the classes of weakly spectrally equivalent networks are larger than the classes of spectrally equivalent networks considered in [3]. Namely each class of equivalence in the weak sense consists of a countable number of equivalence classes in the (strong) sense of [3]. In what follows we will refer to the spectral equivalence in the form introduced in [3] as strong spectral equivalence, and the notion of spectral equivalence introduced in [5] as weak spectral equivalence. Both of the strong and weak notions of spectral equivalence could be of use for analysis of real-world networks many of which have a hierarchical structure [6, 7].

3 Attractors of Isospectral Reductions

Isospectral reductions of networks (graphs) define a dynamical system on the space of all networks. This dynamical system arises by picking any node (edge) of a network and isospectrally reducing this network to a network where the set of nodes is a complement to a chosen node. The fact that such isospectral reductions form a dynamical system follows from the Commutativity theorem proved in [3] which states that a sequence of isospectral compressions over a set of nodes A and then over the set of nodes B gives the same result as isospectral reduction over B followed by the one over A. Therefore to one and the same network (graph) G correspond different orbits depending on the order in which we pick nodes of G for reductions.

By repeatedly compressing a graph in this manner it is possible to isospectrally reduce any network to a trivial network which has just one node, which can be any node of G. It is clearly a senseless operation. However we can choose a reasonable rule which will help us to understand some intrinsic feature(s) of the structure (topology) of the network G. Generally a network can have many different structural sets. To make the isospectral contraction focused on specific properties of networks, we can add some specific rules to the selection of structural sets.

Before we do that, let us recall a few characteristics of nodes in a graph. (There are about ten-fifteen such characteristics of nodes and edges of networks which are all borrowed from the graph theory).

For a graph $G = (V, E, w)$, the indegree for a node $v \in V$, $d^-(v)$, is the number of edges that end in v. The outdegree $d^+(v)$ is the number of edges that start at v.

Let's define $d(v) = d^-(v) + d^+(v)$ to be the sum of the indegree and outdegree for any node.

Let σ_{st} be the total number of shortest paths from node s to node t, and let $\sigma_{st}(v)$ be the number of those paths that pass through v. Note that $\sigma_{st}(v) = 0$ if $v \in \{s, t\}$ or if v does not lie on any shortest path from s to t. We call

$$g(v) = \sum_{s \neq v} \sum_{t \neq v, s} \sigma_{st}(v)$$

the centrality/betweenness of node v.

Theorem 2 *For any network G and a subset selecting rule τ based on some characteristic of its nodes (edges) ($\tau(G) \neq \emptyset$), the orbit of the dynamical system generated by isospectral reductions with respect to τ converges to an attractor which is a network in which τ selects all the nodes (edges).*

Proof If the network is already an attractor, then the reduction doesn't change this network and the orbit is a fixed point.

Otherwise, each reduction removes at least one vertex (edge). Thus an orbit of a network under consecutive isospectral reductions becomes an attractor in no more than N steps, where $N := |V|$ (or $N := |E|$). Therefore an orbit of a finite network G approaches an attractor in a finite number of steps which does not exceed the number of nodes (edges) in G. Such attractor always exists because any network can be isospectrally reduced to a graph with just one node. A process of consecutive isospectral reductions (i.e. an orbit of the corresponding dynamical system) will terminate at one node, if no one of the networks along this orbit was an attractor for τ. Clearly in case of a "network" with only one node (edge) the values of all characteristics of all nodes (edges) are the same because there is only one node (edge). If G is an infinite network then the corresponding orbit could be finite or infinite.

Theorem 3 *The attractors of isospectral reductions with respect to different characteristics of one and the same network are generally different.*

Proof (i) In the example shown in the Fig. 1, all nodes have degree 4. This graph cannot be further reduced based on the degree of its nodes. However, the centrality of the nodes are different. If we count the number of shortest paths through each node, we can see $c(1) = c(2) = c(3) = c(8) = c(9) = c(10) = 1, c(4) = c(6) = c(7) = c(11) = 27, c(5) = 66$. This graph can be further reduced based on centrality. Therefore for this network (graph) attractors with respect to degree and to centrality are different.

(ii) The complete graph, where each and every node and edge have the same properties, can not be further reduced based on degree or other characteristics of a network. It is always an attractor. If we consider isospectral expansion (see [4]) of a complete graph with respect to two different characteristics, then we get two different graphs (networks) with the same attractor with respect to these two characteristics. Clearly this attractor will be the initial complete graph.

Fig. 1 A network which is an attractor with respect to degree but not with respect to centrality

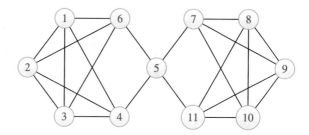

The result of Theorem 3 is not surprising because different characteristics of nodes (or edges) define different dynamical systems on the space of all networks, and orbits of these different dynamical systems are also different.

The following statement establishes that weakly as well as strongly spectrally equivalent networks have the same attractor if isospectral contractions are generated by the very same characteristic with respect to which these networks are spectrally equivalent.

Theorem 4 *Strongly as well as weakly spectrally equivalent graphs with respect to some characteristic have the same attractor under the dynamical system generated by isospectral compressions according to this characteristic.*

Proof Suppose the graph G is strongly spectrally equivalent to H with respect to rule τ, i.e. $R_\tau(G) \simeq R_\tau(H) = R$, and G is weakly spectrally equivalent to K w.r.t τ, i.e. $R^l_\tau(G) \simeq R^m_\tau(K) = S$.

If R is an attractor under τ, then the attractor for G as well as for H is R. So G and H have the same attractor R. Otherwise G and H have the same attractor, the attractor for R. Similarly G and K have the same attractor. Therefore the attractors for all three graphs, G, H, K are the same under rule τ. So all three networks (graphs) have the same attractor with respect to the rule τ.

A very important fact is that networks can be spectrally equivalent with respect to one characteristic of nodes (edges) but not spectrally equivalent with respect to another characteristic. Therefore spectral equivalences built on different characteristics of nodes and edges allow us to uncover various intrinsic (hidden) features of networks' topology.

We now present an example where networks are isomorphic for one characteristic but not for another.

Consider the graphs G and H in Fig. 2.

Their adjacency matrices are

$$M_G = \begin{pmatrix} 1/\lambda & 1 & 1 & 1 & 0 & 0 \\ 0 & 1/\lambda & 1 & 0 & 1 & 0 \\ 0 & 0 & 1/\lambda & 0 & 0 & 1 \\ 1 & 0 & 0 & 0 & 0 & 0 \\ 0 & 1 & 0 & 0 & 0 & 0 \\ 0 & 0 & 1 & 0 & 0 & 0 \end{pmatrix}, \quad M_H = \begin{pmatrix} 2/\lambda & 1 & 1 & 0 & 0 \\ 0 & 1/\lambda & 1 & 1 & 0 \\ 0 & 0 & 1/\lambda & 0 & 1 \\ 0 & 1 & 0 & 0 & 0 \\ 0 & 0 & 1 & 0 & 0 \end{pmatrix}.$$

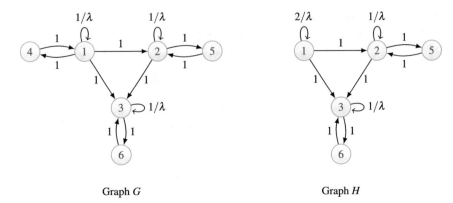

Graph G   Graph H

Fig. 2 Original networks: spectrally equivalent or not?

We can always remove one node in an isospectral reduction. Let us remove node 4 from graph G. The weights of the edges after reduction become

$$R(i, j) = w(i, j) + w(i, 4)\frac{w(4, j)}{\lambda}, \quad i, j = 1, 2, 3, 5, 6.$$

But $w(i, 4) = 0$ for all $i = 2, 3, 5, 6$, and $w(4, j) = 0$ for $j = 2, 3, 5, 6$. The only weight that actually changes after the reduction is $R(1, 1) = w(1, 1) + w(1, 4)w(4, 1)/\lambda = 2/\lambda$. All the other weights satisfy $R(i, j) = w(i, j)$, $i \neq 1$ or $j \neq 1$. The reduced graph after removing node 4 is identical to graph H. Therefore H is an isospectral reduction of G. The networks H and G will have the same reduction as long as we pick the same subset of vertices to reduce on.

We introduce now a few useful notations. For any graph $G = (V, E, w)$, denote the maximum indegree by $m^- = \max\{d^-(v) : v \in V\}$, the maximum outdegree by $m^+ = \max\{d^+(v) : v \in V\}$, and the maximum sum of indegree and outdegree as $m = \max\{d(v) : v \in V\}$. We define a few different rules for picking a subset of the vertices of a graph.

$$\tau_1(G) = \{v \in V : d(v) > m/2\};$$

$$\tau_2(G) = \{v \in V : d^-(v) \geq m^-/2\};$$

$$\tau_3(G) = \{v \in V : d^-(v) > m^-/4\}.$$

The rule τ_1 picks the nodes whose sum of indegree and outdegree is greater than half of the maximum. The rule τ_2 picks the nodes whose indegree is greater than or equal to half of the maximum. And τ_3 picks the nodes whose indegree is greater than a quarter of the maximum.

Now we apply these rules to G and H and see what happens. Consider the degrees of all the nodes in the two graphs. We list them in the following Table 1.

Table 1 The degrees of each node in G and H

Graph	G						H				
Node	1	2	3	4	5	6	1	2	3	5	6
Indegree	2	3	4	1	1	1	1	3	4	1	1
Outdegree	4	3	2	1	1	1	3	3	2	1	1
Sum of indegree and outdegree	6	6	6	2	2	2	4	6	6	2	2

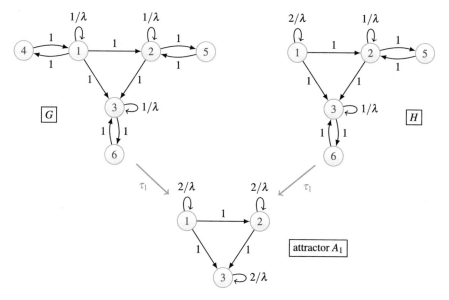

Fig. 3 Isospectral reductions using the rule τ_1

Let us consider τ_1 first. Both G and H have a maximum sum of indegree and outdegree of 6. $\tau_1(G) = \tau_1(H) = \{1, 2, 3\}$. G and H reduce to the same graph in one step under rule τ_1, as shown in Fig. 3. So G and H are spectrally equivalent under the rule τ_1 with respect to both the 1-step definition in [3] and the multi-step definition we have here. Also the reduced graph A_1 is an attractor for the rule τ_1 since the 3 nodes have the same sum of indegree and outdegree, which is 4. To be more precise, if we write down the indegree, outdegree and the sum of the two, (d^-, d^+, d) as an ordered triple for each node, all the triples for the nodes in A_1 are node 1 with $(1, 3, 4)$, node 2 with $(2, 2, 4)$ and node 3 with $(3, 1, 4)$, so $d(1) = d(2) = d(3) = m(A_1)$.

Similarly, for the rule τ_2, we have $\tau_2(G) = \{1, 2, 3\} \neq \tau_2(H) = \{2, 3\}$. However, $\tau_2(\tau_2(G)) = \{2, 3\} = \tau(H)$. Under the rule τ_2, the graph G takes 2 reductions to reach the attractor A_2 while the graph H takes only one step (see Fig. 4). So G and H are spectrally equivalent with our generalized definition but not with respect to the strong definition of spectral equivalence found in [3]. In the graph A_2, the

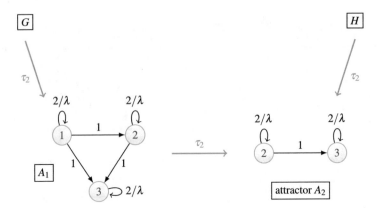

Fig. 4 Isospectral reductions under the rule τ_2

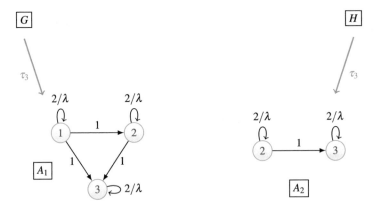

Fig. 5 Isospectral reductions under the rule τ_3

degree triplets for each node are node 2 with $(1, 2, 3)$ and node 3 with $(2, 1, 3)$. Here $d^-(2) = 1 = 1/2m^-(A_2) = 1/2d^-(3)$. One can see A_1 is an attractor of the rule τ_1 but not of the rule τ_2 since $d^-(1) = 1 < 1/2d^-(3) = 3/2$.

Lastly, for τ_3, $\tau_3(G) = \{1, 2, 3\} = \tau_3(\tau_3(G))$, $\tau_3(H) = \{2, 3\} = \tau_3(\tau_3(H))$. Here G and H both reach an attractor in one step. But the attractors they reach are different. Under the rule τ_3 the graphs G and H are not isospectrally equivalent by either definition (see Fig. 5).

Here A_1 and A_2 are both attractors for the rule τ_3. For A_1, $d^-(1) = 1, d^-(2) = 2, d^-(3) = 3$. For A_2 we have $d^-(2) = 1, d^-(3) = 2$. So A_1 is an attractor under the rules τ_1 and τ_3 but not under τ_2. A_2 is an attractor for all 3 rules we used in this sequence of examples.

Theorem 5 *Let $G = (V, E, w)$ with $w : E \to \mathbb{C}$. If S is a structural and $S \subseteq S' \subseteq V$, then S is a structural set of the isospectral reduction $R_{S'}(G)$.*

Proof Suppose $S \subsetneq S' \subsetneq V$. Now we will show that S is also a structural set for the reduced graph $R_{S'}(G)$.

(i) Any cycle (not a loop) in $R_{S'}(G)$ comes from a cycle in G. It has to contain a vertex in S.

(ii) For any $i \in S' \setminus S$, the new weight in $R_{S'}(G)$ is given by

$$\tilde{w}(i, i) = w(i, i) + \sum_{j \in V/S'} w(i, j) \frac{w(j, i)}{\lambda - w(j, j)}$$

$$+ \sum_{j \neq k, j, k \in V/S'} w(i, j) \frac{w(j, k)}{\lambda - w(j, j)} \frac{w(k, i)}{\lambda - w(k, k)} + \dots.$$

Since $w(i, i), w(j, j), w(k, k) \in \mathbb{C}$, the expression above shows that $\tilde{w}(i, i) \neq \lambda$. This implies that S is a structural set of $R_{S'}(G)$.

Remark 3 If we allow the original graph to take weights in $\mathbb{W}$, the above proof still holds as long as $\tilde{w}(i, i) \neq \lambda, \forall i \in S' \setminus S$. Since it's a zero measure set among all the possible values $\tilde{w}(i, i)$'s can take, we can say generally, the theorem is true for any graph with weights in $\mathbb{W}$ except for unusual cases.

By the uniqueness of sequential graph reductions, we can see isospectral reduction is a dynamical system.

Acknowledgements This work was partially supported by the NSF grant CCF-BSF 1615407 and the NIH grant 1RO1EBO25022-01. We would like to thank an anonymous referee for numerous useful comments.

References

1. Newman, M., Barabasi, A.L., Watts, D.J.: The Structure and Dynamics of Networks. Princeton University Press, Princeton (2006)
2. Newman, M.: Networks: An Introduction. Oxford University Press, Oxford (2010)
3. Bunimovich, L., Webb, B.: Isospectral Transformations. Springer, Berlin (2014)
4. Bunimovich, L., Smith, D., Webb, B.: Specialization Models of Network Growth (2017). arXiv:1712.01788
5. Bunimovich, L., Shu, L.: Generalized Eigenvectors of Isospectral Transformations, Spectral Equivalence and Reconstruction of Original Networks (2018). arXiv:1802.03410
6. Clauset, A., Moore, C., Newman, M.: Hierarchical structure and the prediction of missing links in networks. Nature **453**, 98–101 (2008)
7. Leskovec, J., Lang, K., Dasgupta, A., Mahoney, M.: Statistical properties of community structure in large social and information networks. In: Proceedings of the 17th International Conference on World Wide Web, pp. 695–704 (2008)

On Herman's Theorem for Piecewise Smooth Circle Maps with Two Breaks

Akhtam Dzhalilov, Alisher Jalilov and Dieter Mayer

Abstract In this paper we consider general orientation preserving circle homeo-morphisms $f \in C^{2+\varepsilon}(S^1 \setminus \{a^{(0)}, c^{(0)}\})$, $\varepsilon > 0$, with an irrational rotation number ρ_f and two break points $a^{(0)}, c^{(0)}$. Denote by $\sigma_f(x_b) := \frac{Df_-(x_b)}{Df_+(x_b)}$, $x_b = a^{(0)}, c^{(0)}$, the jump ratios of f at the two break points and by $\sigma_f := \sigma_f(a^{(0)}) \cdot \sigma_f(c^{(0)})$ its total jump ratio. Let h be a piecewise-linear (PL) circle homeomorphism with two break points a_0, c_0, irrational rotation number ρ_h and total jump ratio $\sigma_h = 1$. Denote by $\mathbf{B}_n(h)$ the partition determined by the break points of h^{q_n} and by μ_h the unique h-invariant probability measure. It is shown that the derivative Dh^{q_n} is constant on every element of $\mathbf{B}_n(h)$ and takes either two or three values. Furthermore we prove, that $\log Dh^{q_n}$ can be expressed in terms of the μ_h- measures of some intervals of the partition $\mathbf{B}_n(h)$ multiplied by the logarithm of the jump ratio $\sigma_h(a_0)$ of h at the break point a_0. M. Herman showed, that the invariant measure μ_h is absolutely continuous iff the two break points belong to the same orbit. We complement Herman's result for the above class of piecewise $C^{2+\varepsilon}$-circle maps f with irrational rotation number ρ_f and two break points $a^{(0)}, c^{(0)}$ not lying on the same orbit with total jump ratio $\sigma_f = 1$ as follows: if μ_f denotes the invariant measure of the P-homeomorphism f, then for almost all values of $\mu_f([a^{(0)}, c^{(0)}])$ the measure μ_f is singular with respect to Lebesgue measure.

A. Dzhalilov (✉)
Turin Polytechnic University, Kichik Halka yuli 17, Tashkent 100095, Uzbekistan
e-mail: a_dzhalilov@yahoo.com

A. Jalilov
Department of Mathematics, Ajou University, Suwon, South Korea
e-mail: adjalilov@gmail.com

D. Mayer
Institut für Theoretische Physik, TU Clausthal, 38678 Clausthal-Zellerfeld, Germany
e-mail: dieter.mayer@tu-clausthal.de

© Springer Nature Switzerland AG 2018
A. Azamov et al. (eds.), *Differential Equations and Dynamical Systems*,
Springer Proceedings in Mathematics & Statistics 268,
https://doi.org/10.1007/978-3-030-01476-6_7

Keywords Circle homeomorphism · Rotation number · Break point · Invariant measure

1   Introduction

Let f be an orientation preserving homeomorphism of the circle $S^1 \equiv \mathbb{R}/\mathbb{Z}$ with lift $F : \mathbb{R} \to \mathbb{R}$, which is continuous, strictly increasing and fulfills $F(\hat{x} + 1) = F(\hat{x}) + 1$, $\hat{x} \in \mathbb{R}$. The circle homeomorphism g is then defined by $f(x) = F(\hat{x})$ mod 1 with $\hat{x} \in \mathbb{R}$ a lift of $x \in S^1$. The **rotation number** ρ_f is defined by $\rho_f := \lim_{n \to \infty} \frac{F^n(\hat{x}) - \hat{x}}{n}$ mod 1. Here and below, F^i denotes the ith iteration of the map F. It is well known, that the rotation number ρ_f does not depend on the starting point $\hat{x} \in \mathbb{R}$ and is irrational if and only if f has no periodic points (see [5]). The rotation number ρ_f is invariant under topological conjugations.

Denjoy's classical theorem states, that a circle diffeomorphism f with irrational rotation number $\rho = \rho_f$ and $\log Df$ of bounded variation can be conjugated to the linear rotation R_ρ with lift $\hat{R}_\rho(\hat{x}) = \hat{x} + \rho$, that is, there exists a homeomorphism $\varphi : S^1 \to S^1$ with $f = \varphi \circ R_\rho \circ \varphi^{-1}$ [7].

It is well known that a circle homeomorphisms f with irrational rotation number ρ_f is uniquely ergodic, i.e. it has a unique invariant probability measure μ_f. A remarkable fact then is, that the conjugacy φ can be defined by $\varphi(x) = \mu_f([0, x])$ (see [5]), which shows, that the smoothness properties of the conjugacy φ imply corresponding properties of the density of the absolutely continuous invariant measure μ_f for sufficiently smooth circle diffeomorphism with a typical irrational rotation number (see [15, 16]). The problem of smoothness of the conjugacy for smooth diffeomorphisms is by now very well understood (see for instance [3, 14–16, 27]).

A natural generalization of circle diffeomorphisms are piecewise smooth homeomorphisms with break points (see [14]).

The class of **P-homeomorphisms** consists of orientation preserving circle homeomorphisms f which are differentiable except at a finite or countable number of break points, denoted by $BP(f) = \{x_b \in S^1\}$, at which the one-sided positive derivatives Df_- and Df_+ exist, but do not coincide, and for which there exist constants $0 < c_1 < c_2 < \infty$, such that

- $c_1 < Df_-(x_b) < c_2$ and $c_1 < Df_+(x_b) < c_2$;
- $c_1 < Df(x) < c_2$ for all $x \in S^1 \backslash BP(f)$;
- $\log Df$ has finite total variation v in S^1.

Piecewise linear (PL) orientation preserving circle homeomorphisms are simplest examples of P-homeomorphisms. They occur in many other areas of mathematics such as group theory, homotopy theory and logic via the Thompson groups. A family of PL-homeomorphisms were first studied by M. Herman [14] to give examples of circle homeomorphisms of arbitrary irrational rotation number which admit no invariant σ-finite measure absolutely continuous with respect to Lebesque measure. Herman's family of maps has been studied later by several authors (see for instance

[4, 17, 25]) in the context of interval exchange transformations. Special cases are affine 2-interval exchange transformations, to which Herman's examples with break points $a^{(0)} = 0$ and $c^{(0)} = c$ belong.

In [14] Herman proved that the invariant measure of PL-circle homeomorphism with two break points and irrational rotation number is absolutely continuous with respect to Lebesque measure if and only if these break points belong to the same orbit. Invariant measures of more general P-homeomorphisms with one break point have been studied by Dzhalilov and Khanin [9]. In [9] they proved

Theorem 1 *Let $f \in C^{2+\varepsilon}(S^1 \setminus \{a^{(0)}\})$, $\varepsilon > 0$, be a P-homeomorphism with one break point $a^{(0)}$ and irrational rotation number. Then its invariant probability measure μ_f is singular with respect to the Lebesque measure l.*

The invariant measures of P-homeomorphisms f with a finite number of break points have been studied by several authors (see for instance [1, 4, 8–11, 13, 16, 26]). For such a homeomorphism the character of the invariant measure strongly depends on its total jump ratio σ_f being trivial or nontrivial, i.e. $\sigma_f = 1$ or $\sigma_f \neq 1$. A recent result of [13] in the case $\sigma_f \neq 1$ is

Theorem 2 *Let $f \in C^{2+\varepsilon}(S^1 \setminus \{a^{(1)}, a^{(2)}, ..., a^{(m)}\})$, $\varepsilon > 0$ be a P-homeomorphism with irrational rotation number and a finite number of break points $a^{(1)}, a^{(2)}, ..., a^{(m)}$. Suppose its total jump ratio $\sigma_f = \sigma(a^1) \cdot \sigma(a^{(2)}) \cdot ... \sigma(a^{(m)}) \neq 1$. Then its invariant probability measure μ_f is singular with respect to Lebesque measure l.*

More difficult to investigate are piecewise smooth $P-$ homeomorphisms f with a finite number of break points and trivial total jump ratio $\sigma_f = 1$. In the special case of piecewise $C^{2+\varepsilon}$ P-homeomorphisms f, whose break points all lie on the same orbit, the invariant measure μ_f is absolutely continuous w.r.t. to Lebesque measure for typical irrational rotation numbers (see [8]). Rather complicated is the case , when the break points of such a homeomorphism f are not on the same orbit. In this case A. Teplinsky constructed in [26] examples of PL-homeomorphisms f with four break points and trivial total jump ratio $\sigma_f = 1$, whose irrational rotation numbers ρ_f are of unbounded type and whose invariant measures μ_f are absolutely continuous w.r.t. Lebesque measure l. In the present paper we study $C^{2+\epsilon}$ P-homeomorphisms f with arbitrary irrational rotation number ρ_f and two break points not on the same orbit, whose total jump ratio $\sigma_f = 1$. Our main result for these homeomorphisms is

Theorem 3 *Let $f \in C^{2+\varepsilon}(S^1 \setminus \{b^{(1)}, b^{(2)}\})$ be a P-homeomorphism with irrational rotation number $\rho := \rho_f$ and two break points $b^{(1)}, b^{(2)}$ on different orbits with trivial total jump ratio $\sigma_f = \sigma_f(b^{(1)}) \cdot \sigma_f(b^{(2)}) = 1$. Denote its invariant measure by μ_f. Then there exists a subset $M_\rho \subset [0, 1]$ of full Lebesque measure, such that μ_f is singular w.r.t. Lebesgue measure if $\mu_f([b^{(1)}, b^{(2)}]) \in M_\rho$.*

Notice that if $\mu_f([b^{(1)}, b^{(2)}]) = q\rho + p$, $p, q \in Z^1$ then the two break points $b^{(1)}, b^{(2)}$ lie on the same orbit and the invariant measure μ_f is absolutely continuous w.r.t. Lebesgue measure (see [8]).

2   Notations, Terminology, Background

Let f be an orientation preserving circle homeomorphism with irrational rotation number ρ_f. Then ρ_f can be uniquely represented as a continued fraction i.e. $\rho_f = 1/(k_1 + 1/(k_2 + ...)) := [k_1, k_2, ..., k_n, ...]$. Denote by $p_n/q_n = [k_1, k_2, ..., k_n]$, $n \geq 1$, its n-th- convergent. The numbers q_n, $n \geq 1$ are also called the **first return times** of f and satisfy the recurrence relations $q_{n+1} = k_{n+1} q_n + q_{n-1}$ $n \geq 1$, where $q_0 = 1$ and $q_1 = k_1$. Fix an arbitrary $x_0 \in S^1$. Its forward orbit $O_f^+(x_0) = \{x_i = f^i(x_0), \ i = 0, 1, 2...\}$ defines a sequence of natural partitions of the circle. Namely, denote by $I_0^{(n)}(x_0)$ the closed interval in S^1 with endpoints x_0 and $x_{q_n} = f^{q_n}(x_0)$. In the clock-wise orientation of the circle the point x_{q_n} is then for n odd to the left of x_0, and for n even to its right. If $I_i^{(n)}(x_0) = f^i(I_0^{(n)}(x_0))$, $i \geq 1$, denote the iterates of the interval $I_0^{(n)}(x_0)$ under f, it is well known, that the set $\xi_n(x_0)$ of intervals with mutually disjoint interiors, defined as

$$\xi_n(x_0) = \{I_i^{(n-1)}(x_0), \ 0 \leq i < q_n\} \cup \{I_j^{(n)}(x_0), \ 0 \leq j < q_{n-1}\}$$

determines a partition of the circle for any n. The partition $\xi_n(x_0)$ is called the n-th **dynamical partition** of S^1 determined by the point x_0 and the map f. Later we will use also the so called **renormalization intervals** $J_i^{(n)}(x_0) = f^i(J_0^{(n)}(x_0)) = J_0^{(n)}(x_i)$, $i = 0, 1, 2, \ldots$, where $J_0^{(n)}(x_0) = I_0^{(n)}(x_0) \cup I_0^{(n-1)}(x_0)$ and $x_i = f^i(x_0)$.

Proceeding from $\xi_n(x_0)$ to $\xi_{n+1}(x_0)$ all the intervals $I_j^{(n)}(x_0)$, $0 \leq j \leq q_{n-1} - 1$, are preserved, whereas each of the intervals $I_i^{(n-1)}(x_0)$, $0 \leq i \leq q_n - 1$, is partitioned into $k_{n+1} + 1$ subintervals belonging to $\xi_{n+1}(x_0)$, such that

$$I_i^{(n-1)}(x_0) = I_i^{(n+1)}(x_0) \cup \bigcup_{s=0}^{k_{n+1}-1} I_{i+q_{n-1}+sq_n}^{(n)}(x_0).$$

Obviously one has $\xi_1(x_0) \leq \xi_2(x_0) \leq ... \leq \xi_n(x_0) \leq ....$

Definition 1 Let $K > 1$ be a constant. We call two intervals I_1 and I_2 of S^1 K-**comparable**, if the inequality $K^{-1}\ell(I_2) \leq \ell(I_1) \leq K\ell(I_2)$ holds.

Following [15] we recall

Definition 2 An interval $I = [\tau, t] \subset S^1$ is said to be q_n-small, and its endpoints q_n-close, if the intervals $f^i(I)$, $0 \leq i \leq q_n - 1$, are, except for the endpoints, pairwise disjoint.

It follows from the structure of the dynamical partition, that an interval $I = [\tau, t]$ is q_n-small if and only if either $\tau \prec t \preceq f^{q_{n-1}}(\tau)$ or $f^{q_{n-1}}(t) \preceq \tau \prec t$.

Lemma 1 *Let f be a P-homeomorphism with irrational rotation number ρ_f and $|BP(f)| < \infty$. If the interval $I = (x, y) \subset S^1$ is q_n- small and $f^s(x)$, $f^s(y) \notin BP(f)$ for all $0 \le s < q_n$, then for any $k \in [0, q_n)$ Finzi's inequality*

$$e^{-v} \le \frac{Df^k(x)}{Df^k(y)} \le e^v, \tag{1}$$

holds, where v is the total variation of $\log Df$ on S^1.

Proof of Lemma 1. Take any two q_n-close points $x, y \in S^1$ and $0 \le k \le q_n - 1$. Denote by I the open interval with endpoints x and y. Because the intervals $f^i(I)$, $0 \le i \le q_n - 1$ are disjoint, we obtain

$$|\log Df^k(x) - \log Df^k(y)| \le \sum_{j=0}^{k-1} |\log Df(f^j(x)) - \log Df(f^j(y))| \le v,$$

from which inequality (1) follows immediately.

Using Lemma 1 the following lemma can be proven, which plays a key role in the study of the metrical properties of homeomorphisms.

Lemma 2 *Suppose the circle homeomorphism f satisfies the conditions of Lemma 1. Then for any y_0 with $y_s := f^s(y_0) \notin BP(f)$ for all $0 \le s < q_n$, the inequality*

$$e^{-v} \le \prod_{s=0}^{q_n-1} Df(y_s) \le e^v. \tag{2}$$

holds.

Inequality (2) is called **Denjoy's inequality**. It follows from Lemma 2, that the intervals of the dynamical partition $\xi_n(x_0)$ have exponentially small lengths. Indeed one finds

Corollary 1 *Let $I^{(n)}$ be an arbitrary element of the dynamical partition $\xi_n(x_0)$. Then*

$$\ell(I^{(n)}) \le const\ \lambda^n \tag{3}$$

where $\lambda = (1 + e^{-v})^{-\frac{1}{2}} < 1$.

Definition 3 Two homeomorphisms f_1 and f_2 of the circle are said to be topologically equivalent, if there exists a homeomorphism $\varphi : S^1 \to S^1$ such that $\varphi(f_1(x)) = f_2(\varphi(x))$ for any $x \in S^1$.

The homeomorphism φ is called a **conjugacy**. Corollary 1 implies the following generalization of the classical Denjoy theorem:

Theorem 4 *Suppose that a homeomorphism f satisfies the conditions of Lemma 1. Then the homeomorphism f is topologically conjugate to the linear rotation f_ρ.*

Fig. 1 Herman's
PL-homeomorphism

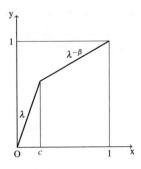

The following fact plays a key role in the proof of Theorems 9–11.

Theorem 5 (see [14], p. 71) *Let f be a P-homeomorphism with irrational rotation number $\rho = \rho_f$ and invariant probability measure μ_f. Then*

$$\int_{S^1} \log Df(x) d\mu_f(x) = 0. \tag{4}$$

3 Herman's Family of PL-Homeomorphisms with Two Break Points

In [14] (Sect. 7 of Chap. VI) M. Herman introduced a family of PL-homeomorphisms with two break points, for which he studied their invariant measures and the regularity of the maps conjugating them to linear rotations: given two real numbers $\lambda > 1$ and $\beta > 0$ he defines for $\hat{x} \in [0, 1]$ the piecewise linear map $F_{\beta,\lambda} : [0, 1] \to [0, 1]$ as

$$F_{\beta,\lambda}(\hat{x}) = \begin{cases} \lambda \hat{x}, & \text{if } 0 \leq \hat{x} \leq c, \\ \lambda^{-\beta}(\hat{x} - 1) + 1, & \text{if } c \leq \hat{x} \leq 1, \end{cases}$$

such that $\lambda c = \lambda^{-\beta}(c - 1) + 1$ (see Fig. 1).

Then Herman considers for $0 \leq \theta \leq 1$ the one-parameter family of PL-maps $F_{\beta,\lambda,\theta}$ of the unit interval with

$$F_{\beta,\lambda,\theta}(\hat{x}) = F_{\beta,\lambda}(\hat{x}) + \theta \mod 1,$$

and the induced piecewise linear homeomorphisms of the circle

$$f_{\beta,\lambda,\theta} = F_{\beta,\lambda,\theta}(\hat{x}) \mod 1. \tag{5}$$

Obviously $a^{(0)} = 0$ and $c^{(0)} = c$ are break points of all these $f_{\beta,\lambda,\theta}$. Denote their rotation number for fixed $\lambda > 1$ and $\beta > 0$ by ρ_θ. Continuity and monotonicity of

ρ_θ as function of θ imply that for arbitrary irrational number $\alpha \in [0, 1]$ there exists an unique $\theta = \theta(\alpha) \in [0, 1]$ with $\rho_\theta = \alpha$. Herman then proved in [14]

Theorem 6 *The following properties are equivalent:*
(i) $f_{\beta,\lambda,\theta}$ is conjugate to the linear rotation f_α through an absolutely continuous homeomorphism;
(ii) $f_{\beta,\lambda,\theta}$ is conjugate to f_α through a Lipschitz homeomorphism;
(iii) $f_{\beta,\lambda,\theta}$ can be conjugated to f_α by a piecewise C^∞ homeomorphism, which is not PL;
(iv) $\frac{\beta}{\beta+1} \in \mathbb{Z}\,\alpha \mod 1$;
(v) the break points $a^{(0)}$ and $c^{(0)}$ belong to the same orbit under $f_{\beta,\lambda,\theta}$.

It can be easily checked (see [14]) that up to two points

$$\frac{\log Df_{\beta,\lambda,\theta}(x)}{(1+\beta)\,\log\lambda} = \chi_{[a^{(0)},c^{(0)}]}(x) - \frac{\beta}{1+\beta}, \tag{6}$$

where $\chi_{[a^{(0)},c^{(0)}]}$ is the characteristic function of the interval $[a^{(0)}, c^{(0)}]$. Obviously

$$\sigma = \sigma(a^{(0)}) = \frac{Df_-(0)}{Df_+(0)} = \lambda^{-1-\beta}$$

for $f = f_{\beta,\lambda,\theta}$, and hence $\log\sigma = -(1+\beta)\log\lambda$. Then we can rewrite (6) as

$$\frac{\log Df_{\beta,\lambda,\theta}(x)}{\log\sigma} = \frac{\beta}{1+\beta} - \chi_{[a^{(0)},c^{(0)}]}(x), \tag{7}$$

and hence also for any $n \geq 1$

$$\frac{\log Df^n_{\beta,\lambda,\theta}(x)}{\log\sigma} = \frac{n\beta}{1+\beta} - \sum_{k=0}^{n-1} \chi_{[a^{(0)},c^{(0)}]}(f^k_{\beta,\lambda,\theta}(x)). \tag{8}$$

Therefore

$$\frac{1}{\log\sigma} \log Df^n_{\beta,\lambda,\theta}(x) = n\,\frac{\beta}{1+\beta} \quad \mod 1, \tag{9}$$

and the following useful Lemma holds for Herman's homeomorphism $f_{\beta,\lambda,\theta}$

Lemma 3 *For every $n \geq 1$*

$$e^{\frac{2\pi i}{\log\sigma} \log Df^n_{\beta,\lambda,\theta}(x)} = e^{2\pi i n \frac{\beta}{1+\beta}}. \tag{10}$$

Furthermore one has

Lemma 4 *Let μ_θ be the invariant measure of Herman's map $f_{\beta,\lambda,\theta}$ with break points $a^{(0)}$, $c^{(0)}$ and irrational rotation number ρ_θ. Then*

$$\mu_\theta([a^{(0)}, c^{(0)}]) = \frac{\beta}{1+\beta}. \tag{11}$$

Proof By Eq. (4) one has $\int_{S^1} \frac{\log Df_{\beta,\lambda,\theta}(x)}{\log \sigma} d\mu_\theta(x) = 0$. Inserting for $\frac{\log Df_{\beta,\lambda,\theta}(x)}{\log \sigma}$ the right side of (7) we get

$$\int_{S^1} \left\{ \frac{\beta}{1+\beta} - \chi_{[a^{(0)}, c^{(0)}0]}(x) \right\} d\mu_\theta(x) = \frac{\beta}{1+\beta} - \mu_\theta([a^{(0)}, c^{(0)}]) = 0,$$

and hence the lemma is proved.

Remark In (11) the right hand side does not depend on parameter θ.
The uniform distribution of sequences is one of the classical problems of ergodic theory (see for instance [21]). Indeed one has

Theorem 7 (see [21]) *For* $[a, b] \subset \mathbb{R}$ *let* $u_n : [a, b] \to \mathbb{R}$, $n = 1, 2, ..$ *be a sequence of continuously differentiable real valued functions. Suppose, for arbitrary* $m, n \in \mathbb{N}$, $n \neq m$, *the function* $Du_n(x) - Du_m(x)$ *is monotone with respect to* x *and that furthermore* $| Du_n(x) - Du_m(x) | \geq K > 0$ *for some constant* K *not depending on* x, m *and* n. *Then the sequence* $u_n(x), n = 1, 2, ...$ *is uniformly distributed* mod 1 *for almost all* x *in* $[a, b]$.

This theorem implies that the sequence $\frac{q_n \cdot \beta}{1+\beta}$ mod 1 is uniformly distributed for almost all β in the sense of Lebesque measure. Clearly, this sequence is not uniformly distributed for all β, since for $\frac{\beta}{1+\beta} = m\rho_{f_{\beta,\lambda,\theta}}$ mod 1 for some integer m, $\lim_{n\to\infty} \| \frac{q_n \cdot \beta}{1+\beta} \| = 0$, where $\| x \|$ denotes the distance of x to the nearest integer. In the case of rotation numbers of bounded type one has the following result.

Theorem 8 (see [22]) *Let* α *be an irrational number of bounded type with partial quotients* $\frac{p_n}{q_n}$. *Then*

$$\lim_{n\to\infty} \| q_n x \| = 0$$

if and only if $x \in \mathbb{Z} \, \alpha$ *mod 1.*

4   On the Location of Break Points

Consider now an arbitrary P-homeomorphism f with irrational rotation number ρ_f and two break points a_0 and c_0, which are not on the same orbit. Denote by $\frac{p_n}{q_n}$ the partial convergents of ρ_f. We will next determine the location of the break points of f^{q_n} and the derivative Df^{q_n} on S^1. Obviously the map f^{q_n} has $2q_n$ break points denoted by $BP_f^n := BP_f^n(a_0) \cup BP_f^n(c_0)$ with $BP_f^n(a_0) := \{a_0^*, a_{-1}^*, ..., a_{-q_n+1}^*\}$, respectively $BP_f^n(c_0) := \{c_0^*, c_{-1}^*, ..., c_{-q_n+1}^*\}$, where $a_{-i}^* = f^{-i}(a_0)$, respectively

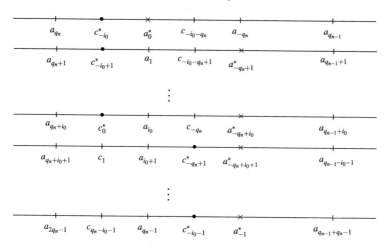

Fig. 2 The position of the break points of f^{q_n} in Lemma 5

$c_{-i}^* = f^{-i}(c_0)$, $0 \le i \le q_n - 1$. It is clear, that these break points of the map f^{q_n} define a partition $B_n(f)$ of the circle S^1 into $2 q_n$ intervals with pairwise non-intersecting interior.

Let $\xi_n(a_0^*)$ be the n-th dynamical partition determined by the break point $a_0^* = a_0$ with respect to the map f. Then one has for the second break point c_0^* either $c_0^* \in I_{i_0}^{(n)}(a_0)$ for some $0 \le i_0 < q_{n-1}$, or $c_0^* \in I_{j_0}^{(n-1)}(a_0) = f^{j_0}((a_0, a_{-q_n}]) \cup f^{j_0}$ $((a_{-q_n}, a_{q_{n-1}}))$ for some $0 \le j_0 < q_n$, i.e. $c_0^* \in f^{j_0}((a_0, a_{-q_n}])$ or $c_0^* \in f^{j_0}((a_{-q_n}, a_{q_{n-1}}))$. The two last cases we have to be treated separately. The following three lemmas describe the location of the break points of f^{q_n} in intervals of certain n-th dynamical partitions.

Lemma 5 *Assume $c_0^* \in I_{i_0}^{(n)}(a_0^*)$ for some i_0 with $0 \le i_0 < q_{n-1}$. Then the break points a_{-i}^*, c_{-i}^*, $0 \le i \le q_n - 1\}$ of f^{q_n} belong to the following elements of the dynamical partition $\xi_n(a_0^*)$ (see Fig. 2):*

- $a_0^* \in I_0^{(n)}(a_0^*)$;
- $c_{-i_0+s}^* = f^s(c_{-i_0}^*) \in I_s^{(n)}(a_0^*)$, $0 \le s \le i_0$;
- $a_{-q_n+s}^* = f^s(a_{-q_n}) \in f^s(a_0^*, a_{-q_n}]) \subset I_s^{(n-1)}(a_0^*)$, $1 \le s \le i_0$;
- $a_{-q_n+s}^*$, $c_{-q_n-i_0+s}^* = f^s(c_{-q_n-i_0}) \in f^s((a_0^*, a_{-q_n}]) \subset I_s^{(n-1)}(a_0^*)$, $i_0 + 1 \le s \le q_n - 1$.

Proof of Lemma 5. Remember, that for arbitrary $x \in S^1$ the points $x_{q_n} = f^{q_n}(x)$ and $x_{-q_n} = f^{-q_n}(x)$ lie on opposite sides of x. Assume $c_0^* = c_0 \in I_{i_0}^n(a_0^*)$ for some $0 \le i_0 < q_{n-1}$ and hence $c_{-i_0}^* \in I_0^n(a_0^*)$ (see Fig. 2). Suppose n to be odd. Then we have in the clockwise order on S^1:

$$a_{q_n} \prec c_{-i_0}^* \prec a_0^* \prec c_{-i_0-q_n} \prec a_{-q_n} \prec a_{q_{n-1}}.$$

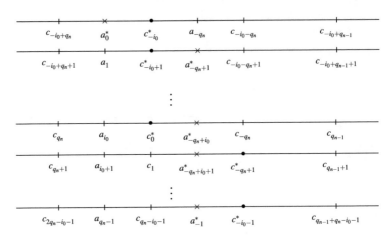

Fig. 3 The position of the break points of f^{q_n} in Lemma 6

Since f is orientation preserving we get also

$$f^s(a_{q_n}) \prec f^s(c^*_{-i_0}) \prec f^s(a^*_0) \prec f^s(c_{-i_0-q_n}) \prec f^s(a_{-q_n}) \prec f^s(a_{q_{n-1}})$$

for all $0 \le s \le i_0$, which proves the first three assertions of Lemma 5.

It is also obvious, that

$$f^s(a^*_0) \prec f^s(c_{-i_0-q_n}) \prec f^s(a_{-q_n}) \prec f^s(a_{q_{n-1}})$$

for all $i_0 + 1 \le s \le q_n - i_0$, which proves the last assertion of Lemma 5.

Lemma 6 *Assume* $c^*_0 \in f^{i_0}((a^*_0, a_{-q_n}])$ *for some* $0 \le i_0 < q_n$. *Then the break points of* f^{q_n} *belong to the following elements of the dynamical partition* $\xi_n(c^*_{-i_0})$ *of the break point* $c^*_{-i_0}$ *(see Fig. 3):*

- $c^*_{-i_0}, a^*_0 \in I_0^{(n)}(c^*_{-i_0})$
- $c^*_{-i_0+s} = f^s(c^*_{-i_0})$, $a^*_{-q_n+s} = f^s(a_{-q_n}) \in f^s([c^*_{-i_0}, a_{-q_n}]) \subset I_s^{(n-1)}(c^*_{-i_0})$, $1 \le s \le i_0$;
- $c^*_{-q_n-i_0+s} = f^s(c^*_{-q_n-i_0})$, $a^*_{-q_n+s} = f^s(a_{-q_n}) \in f^s([c_{-i_0}, c_{-q_n}]) \subset I_s^{(n-1)}(c^*_{-i_0})$, $i_0 + 1 \le s \le q_n - 1$.

Proof of Lemma 6. The interval $J_0^n(c^*_{-i_0}) = [c_{-i_0+q_n}, c_{-i_0+q_{n-1}}]$ contains only the two break points $a^*_0, c^*_{-i_0}$ of f^{q_n}. More precisely, we have (see Fig 3)

$$c_{-i_0+q_n} \prec a^*_0 \prec c^*_{-i_0} \prec a_{-q_n} \prec c_{-i_0-q_n} \prec c_{-i_0+q_{n-1}}$$

which implies the first assertion of Lemma 6.

Next, the renormalization interval $J_1^n(c_0^*) = [c_{-i_0+q_n+1}, c_{-i_0+q_{n-1}+1}]$ contains also two break points of f^{q_n}, namely $a_{-q_n+1}^*$ and $c_{-i_0+1}^*$. The last two break points belong to the interval $I_1^{n-1}(c_{-i_0}^*)$. We have (see Fig. 3)

$$c_{-i_0+1}^* \prec a_{-q_n+1}^* \prec c_{-i_0-q_n+1} \prec c_{-i_0+q_{n-1}+1}.$$

Applying the map f^s for $0 \le s \le i_0 - 1$ leads to

$$f^s(c_{-i_0+1}^*) \prec f^s(a_{-q_n+1}^*) \prec f^s(c_{-i_0-q_n+1}) \prec f^s(c_{-i_0+q_{n-1}+1})$$

which implies the second assertion of Lemma 6.
It is also clear, that

$$c_1 \prec a_{-q_n+i_0+1}^* \prec c_{-q_n+1}^* \prec c_{q_{n-1}+1}$$

Hence for $1 \le s \le q_n - i_0 - 2$

$$f^s(c_1) \prec f^s(a_{-q_n+i_0+1}^*) \prec f^s(c_{-q_n+1}^*) \prec f^s(c_{q_{n-1}+1}).$$

which implies the third assertion of Lemma 6.

Lemma 7 *If* $c_0^* \in f^{i_0}((a_{-q_n}, a_{q_{n-1}}])$ *for some* i_0 *with* $0 \le i_0 < q_n$, *the break points of* f^{q_n} *are located in the following elements of the dynamical partition* $\xi_n(a_{-q_n+1}^*)$ *of the break point* $a_{-q_n+1}^*$ *(see also Fig. 4):*

- $a_{-q_n+1+s}^* = f^s(a_{-q_n+1}^*)$, $c_{-i_0+1+s}^* = f^s(c_{-i_0+1}^*) \in I_s^{(n-1)}(a_{-q_n+1}^*)$, $0 \le s \le i_0 - 1$;
- $a_{-q_n+i_0+1+s}^* = f^s(a_{-q_n+i_0+1}^*)$, $c_{-q_n+1+s}^* = f^s(c_{-q_n+1}^*) \in I_{i_0+s}^{(n-1)}(a_{-q_n+1}^*)$, $0 \le s \le q_n - i_0 - 1$.

Proof of Lemma 7. Consider the n-th dynamical partition $\xi_n(a_{-q_n+1}^*)$ of the break point $a_{-q_n+1}^*$. To determine the location of the break points of f^{q_n} in the intervals of $\xi_n(a_{-q_n+1}^*)$ under the assumption of Lemma 7, we use the structure of this dynamical partition and the monotonicity of f to arrive for $0 \le s \le q_n - 1$ at

$$f^s(a_1) \prec f^s(a_{-q_n+1}^*) \prec f^s(c_{-i_0+1}^*) \prec f^s(a_{q_{n-1}+1}) \prec f^s(c_{-i_0-q_n+1}) \prec f^s(a_{-q_n+q_{n-1}+1}).$$

It is easy to see that the first i_0 of these relations imply the first i_0 claims of Lemma 7 and the last $q_n - i_0$ the remaining ones.

Next we consider a P-homeomorphism f with irrational rotation number ρ_f and two break points $a_0^* := a_0$, $a_{i_0}^* := f^{i_0}(a_0)$, $i_0 > 0$, on the same orbit. Put $n_{i_0} := min\{n : q_n \ge i_0\}$. Assume that $n > n_{i_0}$. If the total jump ratio $\sigma_f = 1$, the map f^{q_n} has $2i_0$ break points

$$a_{-q_n+1}^* := a_{-q_n+1}, a_{-q_n+2}^* := a_{-q_n+2}, \ldots, a_{-q_n+i_0}^* := a_{-q_n+i_0}$$

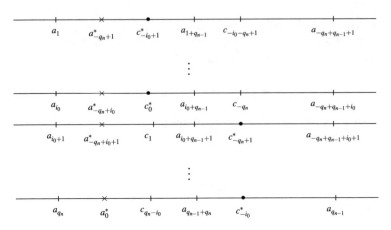

Fig. 4 The position of the break points of f^{q_n} in Lemma 7

Fig. 5 The position of the break points of f^{q_n} in Lemma 8

and

$$a_1^* := a_1, a_2^* := a_2, ..., a_{i_0}^* := a_{i_0}.$$

If $\sigma_f \neq 1$ the map f^{q_n} has $q_n + i_0$ break points

$$a_{-q_n+1}^* := a_{-q_n+1}, \; a_{-q_n+2}^* := a_{-q_n+2}, \; ..., \; a_0^* := a_0, \; ..., \; a_{i_0}^* := a_{i_0}.$$

One has the following

Lemma 8 *Assume f is a P-homeomorphism with irrational rotation number ρ_f and two break points $a_0^* := a_0$, $a_{i_0}^* := f^{i_0}(a_0)$, $i_0 > 0$, on the same orbit. Choose $n > n_{i_0}$.*

(1) If $\sigma_f = 1$, then one finds for the break points $a_{-q_n+s+1}^*$, a_{s+1}^* of f^{q_n}

- $a_{-q_n+s+1}^*, \; a_{s+1}^* \in f^s([a_1^*, a_{-q_n+1}^*]) \subset I_{s+1}^{(n-1)}(a_0^*) \in \xi_n(a_0^*), \; 0 \leq s \leq i_0 - 1$;

(see Fig. 5)

(2) if $\sigma_f \neq 1$, we have

- $a_0^* \subset I_0^{(n-1)}(a_0^*)$;
- $a_{-q_n+1+s}^*, \; a_{1+s}^* \in f^s([a_{i_0+1}, a_{-q_n+i_0+1}^*]) \subset I_{i_0+1+s}^{(n-1)}(a_0^*), \quad 0 \leq s \leq q_n - i_0 - 2.$
- $a_{s+1}^* \in f^s([a_1^*, a_{-q_n+1}^*]) \subset I_{1+s}^{(n-1)}(a_0^*), \quad i_0 \leq s \leq q_n - i_0 - 1.$

Proof of Lemma 8. We will prove the first assertion only. The second one can be proved similarly. It is clear that (see Fig. 5)

$$f^s(a_{q_n+1}) \prec f^s(a_1^*) \prec f^s(a_{-q_n+1}^*) \prec f^s(a_{q_{n-1}} + 1)$$

for all $0 \le s \le i_0 - 1$. Consequently, $f^s(a_1^*)$, $f^s(a_{-q_n+1}^*) \in I_s^{(n-1)}((a_1^*))$, $0 \le s \le i_0 - 1$, which proves the assertion of Lemma 8.

Lemmas 5–8 show the location of the break points of f^{q_n} on elements of different n-th dynamical partitions determined by the map f, respectively their order along the circle. Indeed these lemmas hold true also for any pure rotation f_ρ with ρ_f irrational and any two points $a_0, c_0 \in S^1$, whose preimages under $f_\rho^{q_n}$ correspond to the break points of the P-homeomorphism f^{q_n}.

5 Denjoe Inequality for Piecewise-Linear Circle Homeomorphisms with Two Break Points

Our main goal in this section is to express for a piecewise-linear (PL) homeomorphism h with two break points a_0 and c_0 and total jump ratio $\sigma_h = 1$ the derivative Dh^{q_n} of h^{q_n} by the jump ratio $\sigma_h(a_0)$ and the μ_h-measures of intervals of the partition $B_n(h)$ of S^1 determined by the break points of h^{q_n}.

We apply Lemmas 5–8 to a PL circle homeomorphism h with irrational rotation number ρ_h, with two break points $a_0^* = 0$, $c_0^* = c_0$, not on the same orbit and total jump ratio $\sigma_h = 1$.

In case of Lemmas 5 and 7 the break points $PB_n(a_0^*)$ of h^{q_n} associated with $a_0^* = 0$ and $PB_n(c_0^*)$ associated with $c_0^* = c_0$ alternate in their order along the circle S^1. Let n be odd. Obviously these break points define a system of disjoint subintervals of the circle, given in case of the assumption in Lemma 5 by (see Fig. 2)

$$[c_{-i_0+s}^*, a_{-q_n+s}^*], \quad 1 \le s \le i_0, \tag{12}$$

respectively

$$[c_{-i_0-q_n+s}^*, a_{-q_n+s}^*], \quad i_0 + 1 \le s \le q_n. \tag{13}$$

We combine these subintervals to the subsets

$$A_n(i_0) := \bigcup_{s=1}^{i_0} [c_{-i_0+s}^*, a_{-q_n+s}^*], \quad B_n(i_0) := \bigcup_{s=i_0+1}^{q_n} [c_{-i_0-q_n+s}^*, a_{-q_n+s}^*].$$

In case of the assumption in Lemma 7 the subintervals are given by (see Fig. 4)

$$[a_{-q_n+s}^*, c_{-i_0+s}^*], \quad 1 \le s \le i_0, \tag{14}$$

respectively

$$[a_{-q_n+s}^*, c_{-i_0-q_n+s}^*], \quad i_0 + 1 \le s \le q_n, \tag{15}$$

which we combine to the subsets

$$A_n(i_0) := \bigcup_{s=1}^{i_0} [a^*_{-q_n+s}, c^*_{-i_0+s}], \quad B_n(i_0) := \bigcup_{s=i_0+1}^{q_n} [a^*_{-q_n+s}, c^*_{-i_0-q_n+s}],$$

For n even, the orientation of the above intervals has to be reversed. Therefore in case of Lemma 5 we have the following system of disjoint intervals

$$[a^*_{-q_n+s}, c^*_{-i_0+s}], \quad 1 \le s \le i_0, \tag{16}$$

respectively

$$[a^*_{-q_n+s}, c^*_{-i_0-q_n+s}], \quad i_0+1 \le s \le q_n. \tag{17}$$

In case of Lemma 7 one finds

$$[c^*_{-i_0+s}, a^*_{-q_n+s}], \quad 1 \le s \le i_0, \tag{18}$$

respectively

$$[c^*_{-i_0-q_n+s}, a^*_{-q_n+s}], \quad i_0+1 \le s \le q_n. \tag{19}$$

In case of Lemma 5 and n even, respectively in case of Lemma 7 and n odd, the subsets A_n and B_n can be defined as before. The above constructions show, that the boundaries of every interval in the subsets A_n and B_n consist of break points from $PB_n(a_0^*)$ respectively $PB_n(c_0^*)$. In the following we abbreviate the jump ratio of h at the break point a_0^* by

$$\sigma := \sigma_h(a_0^*) = \frac{Dh_-(0)}{Dh_+(0)}.$$

We can then formulate our first main result.

Theorem 9 *Let h be a PL circle homeomorphism with irrational rotation number ρ_h and two break points $a_0^* = 0$ and $c_0^* := c_0$, whose total jump ratio $\sigma_h = 1$, and which lie on different orbits. Assume c_0^* fulfills the assumptions of Lemma 5 respectively Lemma 7 for some i_0 with $0 \le i_0 < q_{n-1}$. Then in case of Lemma 5*

$$(Dh^{q_n}(x))^{(-1)^n} = \begin{cases} \sigma^{\mu_h(A_n \cup B_n)-1}, & \text{if } x \in A_n \cup B_n \\ \sigma^{\mu_h(A_n \cup B_n)}, & \text{if } x \in S^1 \setminus (A_n \cup B_n); \end{cases} \tag{20}$$

respectively in case of Lemma 7 ,

$$(Dh^{q_n}(x))^{(-1)^{n+1}} = \begin{cases} \sigma^{\mu_h(A_n \cup B_n)-1}, & \text{if } x \in A_n \cup B_n \\ \sigma^{\mu_h(A_n \cup B_n)}, & \text{if } x \in S^1 \setminus (A_n \cup B_n). \end{cases} \tag{21}$$

Theorem 9 shows that Dh^{q_n} is constant on every element of $\mathbf{B}_n(h)$ and takes only two values under the assumptions of Lemmas 5 and 7. Moreover, the values of Dh^{q_n} are determined by the jump ratio $\sigma = \sigma_h(a_0^*)$ and the μ_h-measure of $A_n \cup B_n$.

In case of the assumption on c_0^* in Lemma 6 we can define again a system of disjoint subintervals determined by the elements in $\mathbf{B}_n(h)$. Let n be odd. Then these subintervals are as follows (see Fig. 3):

$$[c_{-i_0+s}^*, a_{-q_n+s}^*], \quad 1 \leq s \leq i_0, \tag{22}$$

respectively

$$[a_{-q_n+s}^*, c_{-i_0-q_n+s}^*], \quad i_0 + 1 \leq s \leq q_n. \tag{23}$$

For n even the orientation of the above intervals has to be reversed. To determine in the case of Lemma 6 the values of Df^{q_n} we define

$$A_n(i_0) := \bigcup_{s=1}^{i_0} [c_{-i_0+s}^*, a_{-q_n+s}^*], \quad B_n(i_0) := \bigcup_{s=i_0+1}^{q_n} [a_{-q_n+s}^*, c_{-i_0-q_n+s}^*]. \tag{24}$$

Then the following theorem holds.

Theorem 10 *Let h be a PL circle homeomorphism with two break points $a_0^* = a_0$ and $c_0^* = c_0$ with $\sigma_h = 1$, which lie on different orbits. Assume c_0^* fulfills the assumption of Lemma 6 for some i_0 with $0 \leq i_0 < q_n$. Then for all $n \geq 1$*

$$(Dh^{q_n}(x))^{(-1)^n} = \begin{cases} \sigma^{\mu_h(A_n)-\mu_h(B_n)-1}, & \text{if } x \in A_n, \\ \sigma^{\mu_h(A_n)-\mu_h(B_n)+1}, & \text{if } x \in B_n, \\ \sigma^{\mu_h(A_n)-\mu_h(B_n)}, & \text{if } x \notin A_n \cup B_n. \end{cases} \tag{25}$$

It remains to discuss the case of a PL-homeomorphism h with irrational rotation number ρ_h and two break points $a_0^* = 0$ and $a_{i_0}^* = h^{i_0}(a_0^*)$, $i_0 > 0$, on the same orbit. In this case the break points of h^{q_n} alternate in their order along the circle S^1. Denote by $U_n(a_s^*)$, $1 \leq s \leq i_0$, the closed intervals with endpoints a_s^* and $a_{-q_n+s}^*$. Obviously these subintervals are disjoint. Lemma 8 implies, that $U_n(a_s^*) \subset I_s^{(n-1)}(a_0^*)$, $1 \leq s \leq i_0$. Next we define for every $n \geq 1$

$$U_n = \bigcup_{s=1}^{i_0} U_n(a_s^*). \tag{26}$$

Then one has

Theorem 11 *Let h be a PL circle homeomorphism with two break points $a_0^* = 0$ and $a_{i_0}^* = h^{i_0}(a_0)$, $i_0 > 0$, with $\sigma_h = 1$, which lie on the same orbit. Put $n_{i_0} := \min\{n : q_n \geq i_0\}$. For $n > n_{i_0}$ one finds*

$$(Dh^{q_n}(x))^{(-1)^{n+1}} = \begin{cases} \sigma^{\mu_h(U_n)}, & \text{if } x \in U_n \\ \sigma^{\mu_h(U_n)-1}, & \text{if } x \in S^1 \setminus U_n, \end{cases} \tag{27}$$

6 Proof of Theorems 9, 10 and 11

Proof of Theorem 9. We prove only the case where c_0^* fulfills the assumptions of Lemma 5. The case with the assumptions of Lemma 7 can be proved analogously. We furthermore restrict ourselves to the case when n is odd. The even case can be handled similarly. Denote the n-th dynamical partition determined by the break point $c_0 = c_0^*$ under the map h by $\eta_n = \eta_n(c_0^*)$. In case c_0^* fulfills the assumption of Lemma 5, we have $a_{q_n} \prec c_{-i_0}^* \prec a_0^* \prec c_{-i_0-q_n} \prec a_{-q_n} \prec a_{q_{n-1}}$. Obviously the function Dh^{q_n} on the circle S^1 is constant on every interval of the partition $\mathbf{B}_n(h)$ determined by all break points of h^{q_n}. It makes jumps determined by the jump ratio $\sigma = \sigma_h(a_0^*)$ at the break points $BP_n(a_0^*)$ and by the jump ratio σ^{-1} at the break points $BP_n(c_0^*)$. By Lemma 5 and taking into account the structure of dynamical partitions it follows that the points of $BP_n(a_0^*)$ and $BP_n(c_0^*)$ alternate in their order around S^1. We "renumerate" all break points of h^{q_n} as follows: $a^{(1)} := a_0^*$ and $c^{(1)} := c_{-i_0}^*$. The other points of $BP_n(a_0^*)$ and $BP_n(c_0^*)$ we denote by $a^{(2)}, a^{(3)}, ..., a^{(q_n)}$ and $c^{(2)}, c^{(3)}, ..., c^{(q_n)}$ in the counterclock direction. Then we have

$$a^{(1)} \prec c^{(q_n)} \prec a^{(q_n)} \prec ... \prec c^{(2)} \prec a^{(2)} \prec c^{(1)} \prec a^{(1)}$$

It is clear that

$$A_n(i_0) \bigcup B_n(i_0) = \bigcup_{s=1}^{q_n} [c^{(s)}, a^{(s)}], \text{ and}$$

$$S^1 \setminus (A_n(i_0) \bigcup B_n(i_0)) = \bigcup_{s=1}^{q_n} (a^{(s+1)}, c^{(s)}) \bigcup (a^{(1)}), c^{(q_n)}).$$

Now we can determine the values of h^{q_n}. For $s > 1$, we have

$$Dh^{q_n}([c^{(s)}, a^{(s)}]) = Dh^{q_n}_-(a^{(s)}) = \sigma Dh^{q_n}_+(a^{(s)}) = \sigma Dh^{q_n}([a^{(s)}, c^{(s-1)}])$$

$$= \sigma Dh^{q_n}_-(c^{(s-1)}) = \sigma \sigma^{-1} Dh^{q_n}_+(c^{(s-1)}) = Dh^{q_n}_+(c^{(s-1)}) = Dh^{q_n}([c^{(s-1)}, a^{(s-1)}])$$

So we get

$$Dh^{q_n}([c^{(s)}, a^{(s)}]) = Dh^{q_n}([c^{(s-1)}, a^{(s-1)}])$$

Iterating the last relation we obtain

$$Dh^{q_n}([c^{(s)}, a^{(s)}]) = Dh^{q_n}([c^{(1)}, a^{(1)}]) \equiv Dh^{q_n}_-(a^{(1)}) = \sigma Dh^{q_n}_+(a^{(1)}) = \sigma Dh^{q_n}_+(a_0^*)$$

Hence Dh^{q_n} takes the constant value $\sigma Dh^{q_n}_+(a_0^*)$ on $A_n(i_0) \bigcup B_n(i_0)$.

Next we show that Dh^{q_n} takes the constant value $Dh_+^{q_n}(a_0^*)$ on $S^1 \setminus (A_n(i_0) \bigcup B_n(i_0))$. First we determine Dh^{q_n} on the interval $[a^{(1)}, c^{(q_n)}]$:

$$Dh^{q_n}([a^{(1)}, c^{(q_n)}]) = Dh_+^{q_n}(a^{(1)}).$$

On the other hand

$$Dh^{q_n}([a^{(s+1)}, c^{(s)}]) = Dh_-^{q_n}(c^{(s-1)}) = \sigma^{-1} Dh_+^{q_n}(c^{(s-1)}) = \sigma^{-1} Dh^{q_n}([c^{(s-1)}, a^{(s-1)}]).$$

This together with

$$Dh^{q_n}([c^{(s-1)}, a^{(s-1)}]) = \sigma Dh_+^{q_n}(a_0^*)$$

implies that for every $s > 1$

$$Dh^{q_n}([a^{(s+1)}, c^{(s)}]) = Dh_+^{q_n}(a_0^*).$$

For the proof of (20) it is enough to prove under the assumption of n being odd and therefore $a_{q_n} \prec c_{-i_0}^* \prec a_0^* \prec c_{-i_0-q_n} \prec a_{-q_n} \prec a_{q_{n-1}}$, that

$$Dh_+^{q_n}(a_0^*) = \sigma^{(-1)^{n+1}\mu_h(U_n)-\delta_{1,(-1)^{n+1}}}, \tag{28}$$

where $\delta_{1,(-1)^{n+1}} = 1$ for n odd, respectively $\delta_{1,(-1)^{n+1}} = 0$ for n even. Notice that the last equation is true also for n even . Since h^{q_n} is an orientation preserving homeomorphism with irrational rotation number and the same invariant measure μ_h as the map h, we get from Theorem 5.

$$\int_{S^1} \log Dh^{q_n}(x) d\mu_h(x) = 0. \tag{29}$$

As mentioned above, the function Dh^{q_n} is constant on the subsets $U_n := A_n(i_0) \cup B_n(i_0)$ and $\overline{U}_n = S^1 \setminus U_n$. Therefore

$$\int_{S^1} \log Dh^{q_n}(x) d\mu_h(x) = \int_{U_n} \log Dh^{q_n}(x) d\mu_h(x) + \int_{\overline{U}_n} \log Dh^{q_n}(x) d\mu_h(x) = 0$$

Inserting the constant values of Dh^{q_n} on the sets U_n respectively $\overline{U}_n$ one finds

$$\int_{U_n} \log Dh^{q_n}(x) d\mu_h = \mu_h(U_n) \log(\sigma Dh_+^{q_n}(a_0^*)),$$

$$\int_{\overline{U}_n} \log Dh^{q_n}(x) d\mu_h = \mu_h(\overline{U}_n) \log Dh_+^{q_n}(a_0^*) = [1 - \mu_f(U_n)] \log Dh_+^{q_n}(a_0^*),$$

and therefore

$$\mu_h(U_n) \log(\sigma \, Dh_+^{q_n}(a_0^*)) + [1 - \mu_h(U_n)] \log Dh_+^{q_n}(a_0^*) = 0.$$

This shows that $\mu_h(U_n) \log \sigma = -\log Dh_+^{q_n}(a_0^*)$ respectively $Dh_+^{q_n}(a_0^*) = \sigma^{-\mu_h(U_n)}$, and hence formula (28) holds for n odd. For n even, the proof of formula (28) proceeds similarly. Theorem 9 is therefore completely proved.

Proof of Theorem 10. We will prove only the following equation

$$Dh_+^{q_n}(a_0^*) = \sigma^{(-1)^n (\mu_h(A_n(i_0)) - \mu_h(B_n(i_0))) - \delta_{1,(-1)^{n-1}}} \tag{30}$$

for n odd. Since the rotation number ρ_h of h is irrational and its break points a_0^* and c_0^* are on different orbits, all the intervals in A_n and B_n are pairwise disjoint. For all $x \in B_n$ one has obviously $Dh^{q_n}(x) = Dh_+^{q_n}(a_0^*)$. But at the break point $c_{i_0}^*$ the function $Dh^{q_n}(x)$ makes the jump $Dh_+^{q_n}(c_{i_0}^*)/Dh_-^{q_n}(c_{i_0}^*) = Dh_+(c_{i_0}^*)/Dh_-(c_{i_0}^*) = \sigma$, and therefore it takes the constant value $Dh^{q_n}(x) = \sigma \, Dh_+^{q_n}(a_0^*)$ in this interval containing no break point of h^{q_n}. Indeed, this holds true for all intervals without break points, i.e. for $x \notin A_n \cup B_n$. The left boundary point of any interval in A_n belongs to the set $BP(c_0^*)$ and hence the function $Dh^{q_n}(x)$ makes at these break points the jump $Dh_+(c_{i_0}^*)/Dh_-(c_{i_0}^*) = \sigma$ and therefore takes the constant value $Dh^{q_n}(x) = \sigma^2 \, Dh_+^{q_n}(a_0^*)$ for any $x \in A_n$. This proves assertion (25).

To prove assertion (30) we use again

$$\int_{S^1} \log Dh^{q_n}(x) d\mu_h(x) = 0,$$

and the possible values of the function Dh^{q_n} discussed above. Then

$$\log(\sigma^2 \, Dh_+^{q_n}(a_0^*))\mu_f h(A_n) + \log(Dh_+^{q_n}(a_0^*))\mu(B_n) + \log(\sigma \, Dh_+^{q_n}(a_0^*))\mu(\overline{U}_n^*) = 0,$$

where $\overline{U}_n^* = S^1 \setminus (A_n \cup B_n)$. Hence

$$(\log \sigma)\{\mu_h(A_n) - \mu_h(B_n)\} + \log Dh_+^{q_n}(a_0^*)) + \log \sigma = 0$$

This proves Eq. (30) for n odd. The proof of the theorem for n even is similar. Theorem 10 is therefore completely proved.

Proof of Theorem 11. Let h be a PL circle homeomorphism with two break points a_0^* and $a_{i_0}^* = f^{i_0}(a_0)$, $i_0 > 0$, and irrational rotation number ρ_h. Assume $n > n_0$. Then h has $2i_0$ break points. Put $BP_h^n := BP_h^n(a_1^*) \cup BP_h^n(a_{-q_n+1}^*)$ with

$$BP_h^n(a_1^*) = \{a_1^*, a_2^*, \ldots, a_{i_0}^*\},$$

respectively

$$BP_h^n(a_{-q_n+1}^*) = \{a_{-q_n+1}^*, a_{-q_n+2}^*, ..., a_{-q_n+i_0}^*\},$$

where $a_s^* = f^s(a_0)$, $a_{-q_n+s}^* = f^s(a_{-q_n})$, $1 \le s \le i_0$. For the proof of Theorem 11 it is sufficient to prove the following formula

$$Dh_+^{q_n}(a_0^*) = \sigma^{(-1)^{n+1}\mu_h(U_n) - \delta_{1,(-1)^{n+1}}} \tag{31}$$

The partition $\mathbf{B}_n(h)$ determined by all break points of h^{q_n} has $2i_0$ closed intervals with disjoint interior. The map Dh^{q_n} is piecewise constant with constant values on the element of $\mathbf{B}_n(h)$. The first assertion of Lemma 8 implies that the intervals in $U_n = \{[a_s^*, a_{-q_n+s}^*], 1 \le s \le i_0\}$ are pairwise disjoint. Hence the intervals of $\overline{U}_n = S^1 \setminus U_n$ are also pairwise disjoint. Next we conclude that

- the break points of $BP_h^n(a_1^*)$ and $BP_h^n(a_{-q_n+1}^*)$ alternate in their order on S^1;
- the intervals in U_n and $\overline{U}_n$ alternate in their order on S^1; Denote by $\overline{U}_n(a_s^*)$ the closed interval in $\overline{U}_n$ with right endpoint a_s^*, $1 \le s \le i_0$. It is easy to see that at each point a_s^* of $BP_h^n(a_1^*)$

$$Dh_+^{q_n}(a_s^*) = Dh_+^{q_n}(a_0^*), \quad Dh_-^{q_n}(a_s^*) = Dh_-^{q_n}(a_0^*), \quad 1 \le s \le i_0. \tag{32}$$

It is clear that the intervals $U_n(a_s^*)$ and $\overline{U}_n(a_s^*)$ are neighbours with common endpoint a_s^*. It is obvious that

$$\frac{Dh_-^{q_n}(a_s^*)}{Dh_+^{q_n}(a_s^*)} = \sigma_h(a_0^*) = \sigma, \quad 1 \le s \le i_0.$$

The last relation together with (32) implies $Dh^{q_n}(x) = \sigma Dh_+^{q_n}(a_0^*)$ if $x \in U_n$ respectively $Dh^{q_n}(x) = Dh_+^{q_n}(a_0^*)$ if $x \in S^1 \setminus U_n$. Remains to determine the value of $Dh_+^{q_n}(a_0^*)$. From Theorem 2.8 we obtain

$$\int_{S^1} \log Dh^{q_n}(x)d\mu_h = \int_{U_n} \log Dh^{q_n}(x)d\mu_h + \int_{\overline{U}_n} \log Dh^{q_n}(x)d\mu_h = 0.$$

Hence $\mu_h(U_n) \log Dh_+^{q_n}(a_0^*) + \mu_h(\overline{U}_n) \log Dh_-^{q_n}(a_0^*) = 0$.

Inserting $\mu_h(\overline{U}_n) = 1 - \mu_h(U_n)$ respectively $Dh_-^{q_n}(a_0^*) = \sigma_h Dh_+^{q_n}(a_0^*)$ we get relation (31). Theorem 11 hence is proved.

7　Proof of Theorem 3

Let $f \in C^{2+\epsilon}(S^1 \setminus \{b^{(1)}, b^{(2)}\})$ be a P-homeomorphism of the circle with irrational rotation number ρ_f and two break points $b^{(1)}$ and $b^{(2)}$ not on the same orbit, whose total jump $\sigma_f = \sigma_f(b^{(1)}) \cdot \sigma_f(b^{(2)}) = 1$. Denote by μ_f its unique invariant probability measure. Define the parameters β and λ through

$$\mu_f([b^{(1)}, b^{(2)}]) =: \frac{\beta}{1+\beta}, \quad \sigma_f(b^{(1)}) =: \lambda^{-1-\beta}. \tag{33}$$

Let $h = h_{\beta,\lambda,\theta}$ be Herman's PL-homeomorphism of S^1 with break points $a^{(0)} = 0$ and $c^{(0)} = c$ such that $\lambda c = \lambda^{-\beta}(c-1) + 1$. Since the rotation number ρ_f is irrational, we can find an unique $\bar{\theta}$ such that the rotation number $\rho_{\bar{\theta}}$ of $h_{\bar{\theta}} = h_{\beta,\lambda,\bar{\theta}}$ coincides with ρ_f. Denote by $\mu_{\bar{\theta}}$ the invariant measure of $h_{\bar{\theta}}$. By Lemma 4 $\mu_{\bar{\theta}}([a^{(0)}, c^{(0)}]) = \frac{\beta}{1+\beta}$. Since $\rho_f = \rho_{\bar{\theta}}$ the homeomorphisms f and $h_{\bar{\theta}}$ are topologically conjugate via some homeomorphism φ. We can choose φ such that $a^{(0)} = \varphi(b^{(1)})$ and $c^{(0)} = \varphi(b^{(2)})$,because $\mu_f[b^{(1)}, b^{(2)}] = \mu_{\bar{\theta}}([a^{(0)}, c^{(0)}])$. Then one has also $(\varphi^* \mu_{\bar{\theta}})([b^{(1)}, b^{(2)}]) = \mu_f[b^{(1)}, b^{(2)}] = \mu_{\bar{\theta}}([a^{(0)}, c^{(0)}])$, since the invariant probability measure of f is unique. Hence we proved the following fact, which will play a key role in our proof of the main Theorem.

Theorem 12 *The P-homeomorphism $f : S^1 \to S^1$ with irrational rotation number ρ_f and two break points $b^{(1)}, b^{(2)}$ with total jump ratio $\sigma_f(b^{(1)}) \cdot \sigma_f(b^{(2)}) = 1$. Let $h_{\beta,\lambda,\bar{\theta}}$ be Herman's PL-homeomorphism with rotation number $\rho_{\bar{\theta}} = \rho_f$ and two break points $a^{(0)}, c^{(0)}$, such that $\sigma_h(a^{(0)}) = \sigma_f(b^{(1)})$ and $\mu_{\bar{\theta}}([a^{(0)}, c^{(0)}]) = \mu_f([b^{(1)}, b^{(2)}])$. Then the maps f and $h_{\beta,\lambda,\bar{\theta}}$ are topologically conjugate by some homeomorphism $\varphi : S^1 \to S^1$ with $\varphi(b^{(1)}) = a^{(0)}$ and $\varphi(b^{(2)}) = c^{(0)}$.*

Since the rotation number ρ_f is irrational, the invariant probability measure μ_f has no discrete ergodic component. Indeed, one knows, that every such P-homeomorphism is ergodic also w.r.t. Lebesque measure l (see [14]). Suppose, μ_f has an absolutely continuous component $\mu_f^{a.c}$ with support A and $\mu_f^{a.c}(A) > 0$. Then also $l(A) > 0$. If $p(x)$ is the density of $\mu_f^{a.c}$, then on A obviously $p(x) \geq 0$ and on $S^1 \setminus A$ one has $p(x) = 0$. Since $p(x)$ satisfies the functional equation $p(f(x)) = \frac{1}{Df(x)} p(x)$, $x \in S^1$ and $Df(x) \geq const > 0$, the subset $A_+ = \{x : p(x) > 0\}$ is f-invariant. Ergodicity of f with respect to Lebesgue measure l implies that either $\mu_f^{a.c}(A) = 1$ or $\mu_f^{a.c}(A) = 0$. Hence the invariant measure μ_f is either pure absolutely continuous or pure singular on S^1.

The main idea of the proof of Theorem 3 is to construct for the homeomorphism f a sequence of measurable subsets $G_{n_m} \subset S^1$, such that $\lim_{m \to \infty} l(G_{n_m}) = \omega \in (0, 1)$ and $\lim_{m \to \infty} | Dh^{q_{n_m}}(x) - 1 | > K > 0$ for all $x \in G_{n_m}$;
– under the assumption, that the invariant measure μ_f is absolutely continuous w.r.t. Lebesque measure l, one knows on the other hand, that $Df^{q_n}(x)$ tends for $n \to \infty$ to 1 in probability with respect to the normalised measure l, which shows μ_f is singular w.r.t. Lebesgue measure;
– since the P-homeomorphism f in Theorem 3 can be conjugated to Herman's map $h_{\bar{\theta}} = h_{\beta,\lambda,\bar{\theta}}$, the structure of the break points of f^{n_m} is identical to the one of these points of $h_{\bar{\theta}}^{q_{n_m}}$. One can then apply a slightly extended reasoning to the distribution of the values of Df^{q_n} relating them to those of $Dh_{\bar{\theta}}^{q_n}$ with respect to the intervals of the partition η_n defined by the breakpoints of $h_{\bar{\theta}}$.

Let us start with the following proposition shown in [10].

Proposition 1 *Let f be a P-homeomorphism of the circle with irrational rotation number ρ_f. If its invariant probability measure μ_f is equivalent to Lebesque measure l, then for all $\delta > 0$ the sequence $l(\{x : | Df^{q_n}(x) - 1 | > \delta\})$ tends to zero as $n \to \infty$.*

Important for our discussion will be also

Lemma 9 *For arbitrary $\delta \in (0, 1)$ and $n \geq 1$ consider three points $z_1, z_2, z_3 \in S^1$ with $z_1 \prec z_2 \prec z_3 \prec z_1$, such that the intervals $[z_1, z_2]$ and $[z_2, z_3]$ are q_n-small. Assume, the P-homeomorphism $f^{q_n} \in C^{2+\varepsilon}(S^1 \setminus \{z_2\})$ has jump ratio $\sigma_{f^{q_n}}(z_2) = \Lambda$ at the break point z_2. For v the total variation of $\log Df$ on S^1 and $t_l \in (z_1, z_2)$ and $t_r \in (z_2, z_3)$ with*

$$\frac{l([t_l, z_2])}{l([z_1, z_2])} = \frac{l([t_r, z_3])}{l([z_2, z_3])} = \delta \tag{34}$$

one has either

$$\log Df^{q_n}(x) \leq -\frac{\log \Lambda}{2} + Ke^v\delta, \tag{35}$$

for all $x \in [t_l, z_2)$, or

$$\log Df^{q_n}(y) \geq \frac{\log \Lambda}{2} - Ke^v\delta \tag{36}$$

for all $y \in (z_2, t_r]$, when $\Lambda > 1$.

In the case $\Lambda < 1$ one has either

$$\log Df^{q_n}(x) \geq -\frac{\log \Lambda}{2} - Ke^v\delta, \tag{37}$$

for all $x \in [t_l, z_2)$, or

$$\log Df^{q_n}(y) \leq \frac{\log \Lambda}{2} + Ke^v\delta \tag{38}$$

for all $y \in (z_2, t_r]$.

Proof Assume $\log \Lambda = \log \frac{Df^{q_n}_-(z_2)}{Df^{q_n}_+(z_2)} > 0$, the case $\log \Lambda < 0$ can be treated analogously. Then $\log Df^{q_n}_-(z_2) = \log \Lambda + \log Df^{q_n}_+(z_2)$, and hence

$$\log Df^{q_n}_+(z_2) \leq -\frac{\log \Lambda}{2} \quad \text{if and only if} \quad \log Df^{q_n}_-(z_2) \leq \frac{\log \Lambda}{2}, \tag{39}$$

respectively

$$\log Df^{q_n}_+(z_2) \geq -\frac{\log \Lambda}{2} \quad \text{if and only if} \quad \log Df^{q_n}_-(z_2) \geq \frac{\log \Lambda}{2}. \tag{40}$$

Hence, either

$$\log Df^{q_n}_+(z_2) \leq -\frac{\log \Lambda}{2} \tag{41}$$

or

$$\log Df_-^{q_n}(z_2) \geq \frac{\log \Lambda}{2}. \tag{42}$$

Then for an arbitrary $x \in [t_l, z_2)$ one finds

$$|\log \frac{Df_-^{q_n}(z_2)}{Df^{q_n}(x)}| \leq \sum_{j=0}^{q_n-1} |\log Df_-(f^j(z_2)) - \log Df(f^j(x))| \quad \leq K \sum_{j=0}^{q_n-1} l([f^j(x), f^j(z_2)])$$

$$\leq K \sum_{j=0}^{q_n-1} l([f^j(t_l), f^j(z_2)]) \leq K \sum_{j=0}^{q_n-1} \frac{l([f^j(t_l), g^j(z_2)])}{l([f^j(z_1), f^j(z_2)])} l([f^j(z_1), f^j(z_2)])$$

$$= K \sum_{j=0}^{q_n-1} \frac{Df^j(\zeta)}{Df^j(\vartheta)} \frac{l([t_l, z_2)])}{l([z_1, z_2)])} l([f^j(z_1), f^j(z_2)]) \text{ for certain } \zeta \in [t^l, z_2), \vartheta \in [z_1, z_2) \text{ and}$$

a universal constant $K > 0$ depending only on f. According to Lemma 1

$$e^{-v} \leq \frac{Df^j(\zeta)}{Df^j(\vartheta)} \leq e^v$$

and therefore

$$K \sum_{j=0}^{q_n-1} \frac{Df^j(\zeta)}{Df^j(\vartheta)} \frac{l([t_l, z_2)])}{l([z_1, z_2)])} l([g^j(z_1), f^j(z_2)]) \leq K e^v \delta,$$

since $\frac{l([t_l, z_2)])}{l([z_1, z_2)])} = \delta$. We used also, that the interval $[z_1, z_2]$ is q_n- small and hence the intervals $[f^j(z_1), f^j(z_2)], 0 \leq j \leq q_n - 1$ are disjoint. This leads finally to the bound

$$|\log \frac{Df_-^{q_n}(z_2)}{Df^{q_n}(x)}| \leq K e^v \delta. \tag{43}$$

In the same way it can be shown that

$$|log \frac{Df_+^{q_n}(z_2)}{Df^{q_n}(y)}| \leq K e^v \delta \tag{44}$$

for all $y \in [t_r, z_2)$. Inserting the bounds (41)–(44) we get the bounds (35)–(38) in Lemma 9.

An important role in the proof of Theorem 3 play certain neighbourhoods of the break points of the P-homeomorphisms f, which we define next. As in the case of Herman's map h_θ denote for fixed $n \geq 1$ by η_n the partition of the circle generated by the $2 q_n$ break points BP_f^n of f^{q_n}, whose elements are the closed intervals with two neighbouring break points of f^{q_n} as boundary points. If $z \in BP_f^n$, we denote by $V_n^l(z)$ respectively $V_n^r(z)$ the interval in η_n whose right respectively left boundary point is the break point z. Given some $\delta \in (0, 1)$, because of Lemma 9 we can then construct left and right subintervals $V_n^l(z; \delta) \subset V_n^l(z)$ respectively $V_n^r(z; \delta) \subset V_n^r(z)$, both with the break point z as a boundary point, such that

$$\frac{l(V_n^l(z;\delta))}{l(V_n^l(z))} = \delta, \quad \frac{l(V_n^r(z;\delta))}{l(V_n^r(z))} = \delta$$

Definition 4 The subintervals $V_n^l(z;\delta)$, $V_n^r(z;\delta)$, respectively the interval $V_n(z;\delta) = V_n^l(z;\delta) \cup V_n^r(z;\delta)$ are called **the left normalized δ-neighbourhood, the right normalized $\delta-$ neighbourhood** respectively **the normalized $\delta-$ neighbourhood** of the break point z.

After these preparations we can now prove Theorem 3. We consider the homeomorphism f as in Theorem 3. Assume there exists for $\rho = \rho_f = \rho_h$ a set M_ρ as in Theorem 3 such that the invariant measure μ_f is absolutely continuous w.r.t. Lebesque measure l if $\mu_f(b^{(1)})$, $b^{(2)}([a^0, c^{(0)}]) \in M_\rho$. We will show that this leads to a contradiction with Proposition 1. For this we have to use certain properties of the distribution of the function $Dh^{q_{n_m}}$ for subsequences q_{n_m}. By (9) we have

$$\frac{1}{\log \sigma} \log Dh_{\beta,\lambda,\theta}^{q_n}(x) = \frac{q_n \cdot \beta}{1+\beta} \quad \text{mod } 1, \qquad (45)$$

Then the sequence $\{\frac{1}{\log \sigma} \log Dh_{\beta,\lambda,\theta}^{q_n}(x) \mod 1\}$ is uniformly distributed on $[0, 1]$, because the sequence $\{\frac{q_n \cdot \beta}{1+\beta} \mod 1\}$ has this property. Hence we can choose for every $\omega \in (0, 1)$ a subsequence $\{n_m, m = 1, 2, ...\}$ such that

$$\lim_{m \to \infty} \frac{1}{\log \sigma} \log Dh_{\beta,\lambda,\theta}^{q_{n_m}}(x) = \omega \quad \text{mod } 1. \qquad (46)$$

Without loss of generality we can assume that for the subsequence $\{m_n, n = 1, 2, ...\}$ the break point $c^{(0)}$ of $Dh^{q_{n_m}}$ fulfills the assumption of one of the three lemmas, either Lemma 5, or Lemma 6 or Lemma 7. Assume, it fulfills the assumption of Lemma 6. Then the step function $\frac{1}{\log \sigma} \log Dh_{\beta,\lambda,\theta}^{q_{n_m}}(x)$ takes only three values determined by $Dh_+^{q_{n_m}}(a^{(0)})$ and the jump ratio σ (see Lemma 6). Using Eq. (28) we obtain

$$\frac{1}{\log \sigma} \log Dh_+^{q_{n_m}}(\bar{a}_0^{(0)}) = (-1)^{n_m}(\mu_h(A_{n_m}(i_0)) - \mu_h(B_{n_m}(i_0))) \quad \text{mod } 1, \qquad (47)$$

and hence with (46)

$$\lim_{m \to \infty} (-1)^{n_m}(\mu_h(A_{n_m}(i_0)) - \mu_h(B_{n_m}(i_0))) = \omega \quad \text{mod } 1. \qquad (48)$$

W.l.o.g. we can assume

$$\lim_{m \to \infty} (\mu_h(A_{n_m}(i_0)) - \mu_h(B_{n_m}(i_0))) = \omega, \qquad (49)$$

with $\omega \in (0, 1)$. We show that

$$\lim_{m \to \infty} (\mu_h(A_{n_m}(i_0)) \cup B_{n_m}(i_0)) = d_1 \tag{50}$$

with $\omega \le d_1 < 1$. Suppose contrary

$$\overline{\lim_{m \to \infty}} (\mu_h(A_{n_m}(i_0)) \cup B_{n_m}(i_0)) = 1; \tag{51}$$

Then $\lim_{m \to \infty} \mu_h(S^1 \setminus (A_{n_m}(i_0) \cup B_{n_m}(i_0)) = 0$. But every interval in A_n and B_n is covered by an interval in the dynamical partition of rank n (see Fig. 3). Then it can easily be shown, that the intervals $(A_{n_m}(i_0)) \cup B_{n_m}(i_0))$ and $S^1 \setminus (A_{n_m}(i_0)) \cup B_{n_m}(i_0)))$ are C- comparable with some constant $C = C(h) > 1$. This on the other hand implies that the limits of the above two sequences are either both zero, or take both values in $(0, 1)$, contradicting to (51) . Consequently, the relation (50) holds. Then

$$\lim_{m \to \infty} \mu_h(S^1 \setminus (A_{n_m}(i_0)) \cup B_{n_m}(i_0))) = d_2$$

with $d_2 = 1 - d_1 \le 1 - \omega < 1$.

Next we prove the assertion of Theorem 1.3.

We have

$$\lim_{m \to \infty} \mu_f(S^1 \setminus (\varphi^{-1}(A_{n_m}(i_0)) \cup \varphi^{-1}(B_{n_m}(i_0)))) = d_2 \ with \ d_2 = 1 - d_1 \le 1 - \omega < 1. \tag{52}$$

We consider two copies of the unit circle. Suppose on the first circle acts the homeomorphism f and on the second Herman's homeomorphism h. Relation (52), the arrangement of the break points of f^{q_n} and absolutely continuity of the invariant measure μ_h w.r.t. Lebesque measure l imply for sufficiently large n_m the following bounds

$$c_1 \le l(\varphi^{-1}(A_{n_m}(i_0)) \cup \varphi^{-1}(B_{n_m}(i_0))) \le c_2 \tag{53}$$

for some constants $c_1, c_2 \in (0, 1)$, and hence also

$$c_3 \le l(S^1 \setminus (\varphi^{-1}(A_{n_m}(i_0)) \cup \varphi^{-1}(B_{n_m}(i_0)))) \le c_4 \tag{54}$$

for some constants $c_3, c_4 \in (0, 1)$. Consider next for a break point $z \in BP(f^{q_{n_m}})$ the left and right normalized δ-neighbourhoods $V_{n_m}^l(z, \delta)$ respectively $V_{n_m}^r(z, \delta)$. Obviously, one of these two normalized δ- neighbourhoods is covered by an interval in $\varphi^{-1}(A_{n_m}(i_0))$ or in $\varphi^{-1}(B_{n_m}(i_0))$, whereas the other one is covered by an interval $I^{(n_m)}(z) \subset S^1 \setminus (\varphi^{-1}(A_{n_m}(i_0)) \cup \varphi^{-1}(B_{n_m}(i_0)))$ of the partition η_{n_m}. We conclude, that the length l of each of these intervals covering the normalized δ-neighbourhoods by definition is δ^{-1} times the length of the latter ones. Define

$$V_{n_m}^l(\delta) := \bigcup_{z \in BP(h^{q_{n_m}})} V_n^l(z, \delta), \quad V_{n_m}^r(\delta) := \bigcup_{z \in BP(h^{q_{n_m}})} V_{n_m}^r(z, \delta).$$

Using then the definitions of $A_{n_m}(i_0)$, $B_{n_m}(i_0)$ (see also Fig. 3) and the normalized one sided δ-neighbourhoods $V_n^{\cdot}(z, \delta)$, we obtain

$$l(A_{n_m}(i_0) \cap (V_{n_m}^l(\delta) \cup V_{n_m}^r(\delta))) = \delta \cdot l(A_{n_m}(i_0)), \tag{55}$$

$$l(B_{n_m}(i_0) \cap (V_{n_m}^l(\delta) \cup V_{n_m}^r(\delta))) = \delta \cdot l(B_{n_m}(i_0)), \tag{56}$$

consequently,

$$l((A_{n_m}(i_0) \cup B_{n_m}(i_0)) \cap (V_{n_m}^l(\delta) \cup V_{n_m}^r(\delta))) = \delta \cdot l(A_{n_m}(i_0) \cup B_{n_m}(i_0)) \tag{57}$$

From this we can now derive bounds on the values of $Df^{q_{n_m}}$. Put $\delta = \left| \frac{\log \sigma}{aKe^v} \right|$, where $a > \left| \frac{\log \sigma}{Ke^v} \right|$. From relations (35) and (36) of Lemma 9 it follows that **either** on the left **or** on the right normalized δ-neighbourhood of every break point z of $h^{q_{n_m}}$

$$|\log Df^{q_{n_m}}(x)| \geq \left| \frac{(a-2)\log \sigma}{2a} \right|. \tag{58}$$

Denote by $G_{n_m}(\delta)$ the union over $z \in BP(f^{q_{n_m}})$ of all those one-sided normalized δ- neighbourhoods, on which $|\log f^{q_n}(x)| \geq \left| \frac{(a-2)\log \sigma}{2a} \right|$. It is clear that $l(G_{n_m}(\delta)) \geq \max\{\delta\, l(A_{n_m}(i_0) \cup B_{n_m}(i_0), l(A_{n_m}(i_0) \cup B_{n_m}(i_0))\}$. Finally we obtain

$$|\log f^{q_{n_m}}(x)| \geq \left| \frac{(a-2)\log \sigma}{2a} \right|$$

for all $x \in G_{n_m}(\delta)$. But this contradicts convergence of $Df^{q_n(x)}$ to one in probability with respect to normalized Lebesgue measure according to Proposition 1.

Acknowledgements This work has partially been done during the first author's (A.D.) visit in 2016 as a senior associate of ICTP, Italy, and a visit in 2015 to Clausthal University, supported by the Volkswagenstiftung through the Lower Saxony Professorship of D.M.

References

1. Adouani, A., Marzougui, H.: Singular measures for class P-circle homeomorphisms with several break points. Ergod. Theory Dyn. Sys. **34**, 423–456 (2014)
2. Akhadkulov, H., Dzhalilov, A., Mayer, D.: On conjugations of circle homeomorphisms with two break points. Ergod. Theory Dyn. Sys. **34**(3), 725–741 (2014). https://doi.org/10.1017/etds.2012
3. Arnold, V.I.: Small denominators I. Mapping the circle onto itself. Izv. Akad. Nauk SSSR Ser. Mat. **25**, 21–86 (1961)
4. Coelho, Z., Lopes, A., da Rocha, L.: Absolutely continuous invariant measures for a class of affine interval exchange maps. Proc. Amer. Math. Soc. **123**(11), 3533–3542 (1995)

5. Cornfeld, I.P., Fomin, S.V., Sinai, Y.G.: Ergodic Theory. Springer, Berlin (1982)
6. Cunha, K., Smania, D.: Rigidity for piecewise smooth homeomorphisms on the circle. Adv. Math. **250**, 193–226 (2014)
7. Denjoy, A.: Sur les courbes définies par les équations différentielles à la surface du tore. J. de Math. Pures et Appl. **11**(9), 333–375 (1932)
8. Dzhalilov, A.A.: piecewise smoothness of conjugate homeomorphisms of a circle with corners. Theor. Math. Phys. **120**(2), 961–972 (1999) (in Russian) English transl.: Theor. Math. Phys. **120**(2), 179–192 (1999)
9. Dzhalilov, A.A., Khanin, K.M.: On invariant measure for homeomorphisms of a circle with a point of break. Funct. Anal. i Prilozhen. **32**(3), 11–21 (1998), translation in Funct. Anal. Appl. **32**(3), 153–161 (1998)
10. Dzhalilov, A.A., Liousse, I.: Circle homeomorphisms with two break points. Nonlinearity **19**, 1951–1968 (2006)
11. Dzhalilov, A.A., Liousse, I., Mayer, D.: Singular measures of piecewise smooth circle homeomorphisms with two break points. Discrete Contin. Dyn. Syst. **24**(2), 381–403 (2009)
12. Dzhalilov, A.A., Akin, H., Temir, S.: Conjugations between circle maps with a single break point. J. Math. Anal. Appl. **366**, 1–10 (2010)
13. Dzhalilov, A.A., Mayer, D., Safarov, U.A.: Piecwise-smooth circle homeomorphisms with several break points. Izv. Ross. Akad. Nauk Ser. Mat. **76**(1), 101–120 (2012), translation in Izv. Math. **76**(1), 94–112 (2012)
14. Herman, M.: Sur la conjugaison différentiable des difféomorphismes du cercle à des rotations. Inst. Hautes Etudes Sci. Publ. Math. **49**, 5–233 (1979). Ergod. Theory Dyn. Syst. **9**, 681–690 (1989)
15. Katznelson, Y., Ornstein, D.: The absolute continuity of the conjugation of certain diffeomorphisms of the circle
16. Khanin, K.M., Sinai, Ya.G.: Smoothness of conjugacies of diffeomorphisms of the circle with rotations. Russ. Math. Surveys **44**(1), 69–99 (1989), translation of Uspekhi Mat. Nauk **44**(1), 57–82 (1989)
17. Khanin, K.M., Vul, E.B.: Circle homeomorphisms with weak discontinuities. Adv. Soviet Math. **3**, 57–98 (1993)
18. Khanin, K.M., Khmelev, D.: Renormalizations and rigidity theory for circle homeomorphisms with singularities of the break type. Commun. Math. Phys. **235**, 69–124 (2003)
19. Khanin, K., Teplinsky, A.: Robust rigidity for diffeomorphisms with singularities. Invent. Math. **169**, 193–218 (2007)
20. Khanin, K.M., Teplinsky, A.Y.: Herman's theory revisited. Invent. math. **178**, 333–344 (2009)
21. Kuipers, L., Niederreiter, H.: Uniform Distribution of Sequences. Wiley, New York (1974)
22. Larcher, G.: A convergence problem connected with continued fractions. Proc. Am. Math. Soc. **103**, N3 (1988)
23. Liousse, I.: Nombre de rotation, mesures invariantes et ratio set des homéomorphisms affines par morceaux du cercle. Ann. I. Fourier **55**(2), 431–482 (2005)
24. Marmi, S., Moussa, P., Yoccoz, J.C.: Linearization of generalized interval exchange maps. Ann. Math. **176**, 1583–1646 (2012)
25. Nakada, H.: Piecewise linear homeomorphisms of type III and the ergodicity of the cylinder flows. Keio Math. Sem. Rep. N **7**, 29–40 (1982)
26. Teplinsky, A.: A circle diffeomorphism with breaks that is smoothly linearizable
27. Yoccoz, J.C.: Il n'y a pas de contre-exemple de Denjoy analytique. C. R. Acad. Sci. Paris Sér. I Math. **298**(7), 141–144 (1984)

Pursuit Game for an Infinite System of First-Order Differential Equations with Negative Coefficients

Ibragimov Gafurjan, Usman Waziri, Idham Arif Alias and Zarina Bibi Ibrahim

Abstract We consider a pursuit differential game described by an infinite system of 1st-order differential equations with negative coefficients in Hilbert space. The control functions of players are subject to integral constraints. The pursuer attempts to bring the system from a given initial state to another state for a finite time and the evader's purpose is opposite. We obtain a condition of completion of pursuit when the control resource of the pursuer is greater than that of the evader. We study a control problem as well.

Keywords Pursuer · Evader · Infinite system of differential equations · Control strategy

1 Introduction

Differential games in finite dimensional Euclidean spaces were studied by many researchers and developed important methods (see, for instance, [10, 25, 28, 30, 36, 37].)

I. Gafurjan (✉) · I. A. Alias
Department of Mathematics and Institute for Mathematical Research,
Universiti Putra Malaysia, Serdang, Malaysia
e-mail: ibragimov@upm.edu.my

I. A. Alias
e-mail: idham_2@upm.edu.my

U. Waziri · Z. B. Ibrahim
Faculty of Science, Department of Mathematics, Universiti Putra Malaysia,
Serdang, Malaysia
e-mail: usmanwazirimth@yahoo.com

Z. B. Ibrahim
e-mail: zarinabb@upm.edu.my

© Springer Nature Switzerland AG 2018
A. Azamov et al. (eds.), *Differential Equations and Dynamical Systems*,
Springer Proceedings in Mathematics & Statistics 268,
https://doi.org/10.1007/978-3-030-01476-6_8

101

There are mainly two constraints on control functions of players: geometric and integral constraints. In-views of the amount of works been done in developing the differential games, the integral constraints have been extensively discussed by many researchers with various approaches (see, for example, [4, 5, 8, 11, 12, 18–21, 26, 27, 29, 31, 34, 35, 39, 42–44]).

One of the powerful tools in studying the control and differential game problems in systems with distributed parameters is the decomposition method. Using this method the control or differential game problem is reduced to ones described by infinite systems of differential equations (see, for example, [2, 6, 7, 9, 13, 32, 40, 41, 45, 46]). We demonstrate briefly the method for the following parabolic equation

$$\frac{\partial z(x, t)}{\partial t} + Az(x, t) = w(x, t), \quad z(x, 0) = z_0(x), \tag{1}$$

where $0 \leq t \leq T$, T is a given positive number, $x = (x_1, \ldots, x_n) \in \Omega \subset R^n, n \geq 1$, Ω is a bounded set with piecewise smooth boundary,

$$Az = -\sum_{i,j=1}^{n} \frac{\partial}{\partial x_i} \left(a_{ij}(x) \frac{\partial z}{\partial x_j} \right).$$

$a_{ij}(x) = a_{ji}(x)$, $x \in \Omega$, and, for some $c > 0$ and for all

$$(\xi_1, \ldots, \xi_n) \in R^n, x \in \Omega, \sum_{i,j=1}^{n} a_{ij}(x)\xi_i\xi_j \geq c \sum_{i=1}^{n} \xi_i^2.$$

The domain of the operator A is the space of twice continuously differentiable functions with compact support in Ω, denoted by $\overset{\circ}{C}{}^2 (\Omega)$. Define inner product

$$(z, y)_A = (Az, y), \quad z, y \in \overset{\circ}{C}{}^2 (\Omega).$$

Then $\overset{\circ}{C}{}^2 (\Omega)$ becomes incomplete Euclidean space. To obtain a complete Hilbert space associated with the operator A, we complete the space $\overset{\circ}{C}{}^2 (\Omega)$ with respect to the norm $||z||_A = \sqrt{(Az, z)}$, $z \in \overset{\circ}{C}{}^2 (\Omega)$. We use the fact that the operator A has countably many eigenvalues

$$\lambda_1, \lambda_2, \ldots, \quad 0 < \lambda_1 \leq \lambda_2 \leq \ldots, \quad \lim_{k \to \infty} \lambda_k = +\infty,$$

and generalized eigenfunctions $\varphi_1, \varphi_2, \ldots$, which is a complete orthonormal system in $L_2(\Omega)$ [33].

Next, let $C(0, T; H_r(\Omega))$ and $L_2(0, T; H_r(\Omega))$ denote the spaces of continuous and measurable functions defined on $[0, T]$ with the values in

$$H_r(\Omega) = \left\{ f \in L_2(\Omega) \mid f = \sum_{i=1}^{\infty} \alpha_i \varphi_i, \ \sum_{i=1}^{\infty} \lambda_i^r \alpha_i^2 < \infty \right\},$$

respectively, where r is a given number. The space $H_r(\Omega)$ is a Hilbert space with inner product and norm defined as follows: if

$$f = \sum_{i=1}^{\infty} \alpha_i \varphi_i \in H_r(\Omega), \ \ g = \sum_{i=1}^{\infty} \beta_i \varphi_i \in H_r(\Omega),$$

then

$$(f, g) = \sum_{i=1}^{\infty} \lambda_i^r \alpha_i \beta_i, \ \ \|f\| = \left(\sum_{i=1}^{\infty} \lambda_i^r \alpha_i^2 \right)^{1/2}.$$

It was proved [2] that if $w(\cdot) \in L_2(0, T; H_r(\Omega))$, then the initial value problem (1) has a unique solution $z(\cdot) \in C(0, T; H_{r+1}(\Omega))$. Next, represent the functions $z(x, t)$ and $w(x, t)$ as

$$z(x, t) = \sum_{k=1}^{\infty} z_k(t) \varphi_k(x), \ \ w(x, t) = \sum_{k=1}^{\infty} w_k(t) \varphi_k(x), \ \ z_k(\cdot), w_k(\cdot) \in L_2(0, T), \ \ k = 1, 2, \ldots,$$

and substitute them into the Eq. (1), and then equate the coefficients at $\varphi_k(x)$ to obtain

$$\dot{z}_k + \lambda_k z_k = w_k, \ \ z_k(0) = z_{k0}, \ \ k = 1, 2, \ldots,$$

where $w_k, z_k, z_{k0} \in R^1$, $k = 1, 2, \ldots$, w_k, are control parameters, $z_{k0} = (z_0, \varphi_k)$. Thus, we have obtained an infinite system of differential equations. Usually, the control function is subjected to geometric or integral constraint. The geometric and integral constraints for the control function $w \in H(0, T; H_r(\Omega))$ of the form

$$\|w(x, t)\| \le \rho, \ \int_0^T \|w(x, t)\|^2 dt \le \rho^2,$$

respectively, can be written as follows

$$\left(\sum_{k=1}^{\infty} \lambda_k^r w_k^2(t) \right)^{1/2} \le \rho, \ \ \sum_{k=1}^{\infty} \lambda_k^r \int_0^T w_k^2(t) dt \le \rho^2,$$

respectively.

Hence, there is an important connections between control problems described by PDE and those described by infinite system of differential equations. Control and differential game problems described by infinite system of differential equations are of independent interest and can be investigated within one theoretical framework independently of those described by PDE assuming that the coefficients $\lambda_k, k = 1, 2, \ldots$, are any real numbers. Of course, in the case where λ_k are any real numbers, we must give adequate definitions of state space, solution of infinite system of differential equations. Also, we have to prove the existence-uniqueness of solution in the state space.

There are several works devoted to control or differential game problems described by infinite system of differential equations (see, for example, [1, 3, 14, 16, 17, 22–24, 38]).

In the paper [14] a differential game problem described by the following infinite system of differential equations

$$\dot{z}_k + \lambda_k z_k = -u_k + v_k, \ z_k(0) = z_{k0}, \ k = 1, 2, \ldots, \tag{2}$$

where $z_k, u_k, v_k \in \mathbb{R}^1$, and $\lambda_k, k = 1, 2, \ldots$, are positive numbers, was studied when integral constraints are subjected to control functions of the players.

In the present paper, we study a pursuit differential game problems described by (2) in the case of negative coefficients $\lambda_k, k = 1, 2, \ldots$. Pursuer tries to bring the state of the system from an initial state z^0 to another given one z^1 for a finite time. Previous studies of differential games described by infinite system of differential equations have only dealt with the case $z^1 = 0$. We obtain sufficient conditions of completion of pursuit.

2 Statement of Problem

Consider the following Hilbert space

$$l_r^2 = \left\{ \alpha = (\alpha_1, \alpha_2, \ldots) | \sum_{k=1}^{\infty} |\lambda_k|^r \alpha_k^2 < \infty \right\},$$

where, r is a real number and $\lambda_1, \lambda_2, \ldots$, is a bounded sequence of negative numbers, with inner product and norm defined by

$$\langle \alpha, \beta \rangle_r = \sum_{k=1}^{\infty} |\lambda_k|^r \alpha_k \beta_k, \ \alpha, \ \beta \in l_r^2, \ \|\alpha\| = \left(\sum_{k=1}^{\infty} |\lambda_k|^r \alpha_k^2 \right)^{1/2}.$$

Let

$$L_2(0, T, l_r^2) = \left\{ w(\cdot) = (w_1(\cdot), w_2(\cdot), \ldots) | \ \|w(\cdot)\|_{L_2(0, T, l_r^2)} < \infty, \ w_k(\cdot) \in L_2(0, T) \right\},$$

where $T > 0$ is a given sufficiently big number,

$$\|w(\cdot)\|_{L_2(0,T,l_r^2)} = \left(\sum_{k=1}^{\infty} |\lambda_k|^r \int_0^T w_k^2(t)dt \right)^{1/2},$$

We examine control and pursuit differential game problems described by the following infinite system of differential equations

$$\dot{z}_k + \lambda_k z_k = -u_k + v_k, \ z_k(0) = z_k^0, \ k = 1, 2, \ldots, \tag{3}$$

where $z_k, u_k, v_k \in \mathbb{R}^1, \ k = 1, 2, \ldots; u = (u_1, u_2, \ldots)$ is the control parameter of pursuer and $v = (v_1, v_2, \ldots)$ is that of evader, $z^0 = (z_1^0, z_2^0, \ldots) \in l_{r+1}^2$.
    Let

$$S(\rho_0) = \left\{ w(\cdot) \in L_2(0, T, l_r^2) | \ \|w(\cdot)\|_{L_2(0,T,l_r^2)} \le \rho_0 \right\},$$

where ρ_0 is a given positive number.

Definition 1 Functions $w(\cdot) \in S(\rho_0)$, $u(\cdot) \in S(\rho)$, and $v(\cdot) \in S(\sigma)$ are called admissible control, admissible control of pursuer, and admissible control of evader, respectively, where ρ and σ are given positive numbers.

It's assumed that $\rho > \sigma$.

Definition 2 Let $w(\cdot) \in S(\rho_0)$. A function $z(t) = (z_1(t), z_2(t), \ldots)$, $0 \le t \le T$, with $z_k(0) = z_k^0, \ k = 1, 2, \ldots$, is called solution of the initial value problem

$$\dot{z}_k(t) + \lambda_k z_k(t) = w_k(t), \ z_k(0) = z_k^0, \ k = 1, 2, \ldots, \tag{4}$$

if $z_k(t), \ k = 1, 2, \ldots$, are absolutely continuous and almost everywhere on $[0, T]$ satisfy the Eq. (4).

Let $C(0, T; l_{r+1}^2)$ be the space of continuous functions $z(t) = (z_1(t), z_2(t), \ldots) \in l_{r+1}^2$ defined on $[0, T]$. We need the following proposition [15].

Proposition 1 If $w(\cdot) \in S(\rho)$, then infinite system of differential equations (4) has the only solution $z(t) = (z_1(t), z_2(t), \ldots)$, $0 \le t \le T$, in the space $C(0, T; l_{r+1}^2)$, where

$$z_k(t) = e^{\beta_k t} \left(z_k^0 + \int_0^t w_k(s)e^{-\beta_k s}ds \right), \ k = 1, 2, \ldots,$$

with $\beta_k = -\lambda_k > 0$.

Note that this existence-uniqueness theorem for the system (4) was proved for any finite interval $[0, T]$. Therefore, we investigate the system (3) and (4) on $[0, T]$.

Definition 3 A function

$$U(t, v) = (U_1(t, v), U_2(t, v), \ldots), \ U : [0, T] \times l_r^2 \to l_r^2,$$

with the components of the form

$$U_k(t, v) = w_k(t) + v_k(t), \ k = 1, 2, \ldots,$$

is referred to as the strategy of pursuer, if, for any admissible control of evader $v(\cdot) = (v_1(\cdot), v_2(\cdot), \ldots)$, the system (3) has the only solution at $u(t) = U(t, v)$, where $w(\cdot) = (w_1(\cdot), w_2(\cdot), \ldots) \in S(\rho - \sigma)$.

We are given another state $z^1 = (z_1^1, z_2^1, \ldots) \in l_{r+1}^2$.

Definition 4 We say that the game (3) can be completed for the time θ ($\theta \leq T$), if there exists a strategy U of pursuer such that, for any admissible control of evader, $z(\tau) = z^1$ at some time $\tau, 0 \leq \tau \leq \theta$.

Pursuer tries to bring the state of the system (3) from z^0 to z^1, and the purpose of evader is opposite. Formulate the problems.

Problem 1 Find a condition on the states $z^0, z^1 \in l_{r+1}^2$ such that the state $z(t)$ of the system (4) can be transferred from the initial position z^0 to the final position z^1 for a finite time.

Problem 2 Find a condition on the states $z^0, z^1 \in l_{r+1}^2$, for which pursuit can be completed in the game (3) for a finite time.

3   Control Problem

In this section, we study a control problem for transferring the system $z(t)$ from the initial position z^0 to the final position z^1.

For the system (4), we study the control problem: find a time θ such that

$$z(0) = z^0, \ z(\theta) = z^1. \tag{5}$$

First, we analysis the following series

$$E(t) = E_1(t) + E_2(t), \ t > 0, \tag{6}$$

where

$$E_1(t) = 2 \sum_{k=1}^{\infty} \beta_k^r |z_k^0|^2 \phi_k(t), \quad E_2(t) = 2 \sum_{k=1}^{\infty} \beta_k^r |z_k^1|^2 \psi_k(t), \tag{7}$$

$$\phi_k(t) = \frac{2\beta_k}{1 - e^{-2\beta_k t}}, \quad \psi_k(t) = \frac{2\beta_k}{e^{2\beta_k t} - 1}, \ k = 1, 2, \ldots.$$

Lemma 1 Let $z^0, z^1 \in l_{r+1}^2$. If, in addition, $z^0, z^1 \in l_r^2$, then the series $E(t)$ converges at any $t > 0$.

Proof Let $z^0, z^1 \in l_r^2$. To show that the series (6) converges, we show that the series $E_1(t)$ and $E_2(t)$ converge. Since β_k is a bounded sequence of positive numbers, therefore $\beta = \sup\limits_k \beta_k < \infty$. Since $\beta_k \leq \beta$, then it is not difficult to show that

$$\phi_k(t) = \frac{2\beta_k}{1 - e^{-2\beta_k t}} \leq \frac{2\beta}{1 - e^{-2\beta t}},$$

which implies that

$$E_1(t) \leq \frac{4\beta}{1 - e^{-2\beta t}} \sum_{k=1}^{\infty} \beta_k^r |z_k^0|^2.$$

The series on the right hand side of this inequality is convergent since $z^0 \in l_r^2$. Thus, the series $E_1(t)$ is convergent.

We can see that $\psi_k(t) \leq \frac{1}{t}, t > 0, k = 1, 2, \ldots.$ Then

$$E_2(t) \leq \frac{2}{t} \sum_{k=1}^{\infty} \beta_k^r |z_k^1|^2.$$

The series on the right hand side of this inequality is convergent since $z^1 \in l_r^2$. Thus, the series $E_2(t)$ is convergent. This completes the proof of Lemma 1.

We'll need some properties of $E(t)$.

Property 1 *$E(t)$ has the following properties:*

(i) $E(t)$ is decreasing on $(0, +\infty)$;
(ii) $E(t) \to +\infty$ as $t \to 0^+$;
(iii) $E(t) \to 4 \sum\limits_{k=1}^{\infty} \beta_k^{r+1} |z_k^0|^2$ as $t \to +\infty$.

Proof The first property follows from the fact that $\psi_k(t)$ and $\phi_k(t), k = 1, 2, \ldots$, are decreasing on $(0, +\infty)$.

The proof of the property (ii) follows from the observations that $\psi_k(t) \to +\infty$ and $\phi_k(t) \to +\infty$, as $t \to 0^+$ for each k.

Finally, we prove the property (iii). According to Lemma 1, $E(t)$ is convergent for any $t > 0$. We fix $t_0 > 0$. Since $E(t_0)$ is convergent, then for any $\varepsilon > 0$, there exists a positive integer N such that

$$F(t_0) = \sum_{k=N+1}^{\infty} \beta_k^r \left(2|z_k^0|^2 \phi_k(t_0) + 2|z_k^1|^2 \psi_k(t_0)\right) < \frac{\varepsilon}{3}, \tag{8}$$

and also

$$\sum_{k=N+1}^{\infty} 4\beta_k^{r+1} |z_k^0|^2 < \frac{\varepsilon}{3} \tag{9}$$

since $z^0 \in l^2_{r+1}$. Then, $F(t) < \frac{\varepsilon}{3}$ for all $t \geq t_0$ since the functions $\psi_k(t)$ and $\phi_k(t)$ are decreasing on $(0, +\infty)$ for each k.

On the other hand, there exists number $T_1 > 0$ such that, for all $t > T_1$,

$$\left| 2 \sum_{k=1}^{N} \beta_k^r \left(|z_k^0|^2 \phi_k(t) + |z_k^1|^2 \psi_k(t) \right) - 4 \sum_{k=1}^{N} \beta_k^{r+1} |z_k^0|^2 \right| < \frac{\varepsilon}{3}, \tag{10}$$

since the sum consists of a finite number of summands and

$$\lim_{t \to +\infty} \phi_k(t) = 2\beta_k, \quad \lim_{t \to +\infty} \psi_k(t) = 0, \quad k = 1, 2, \ldots$$

Thus, by (8)–(10)

$$\left| E(t) - 4 \sum_{k=1}^{\infty} \beta_k^{r+1} |z_k^0|^2 \right| \leq \left| 2 \sum_{k=1}^{N} \beta_k^r \left(|z_k^0|^2 \phi_k(t) + |z_k^1|^2 \psi_k(t) \right) - 4 \sum_{k=1}^{N} \beta_k^{r+1} |z_k^0|^2 \right|$$

$$+ 2 \sum_{k=N+1}^{\infty} \beta_k^r \left(|z_k^0|^2 \phi_k(t) + |z_k^1|^2 \psi_k(t) \right) + 4 \sum_{k=N+1}^{\infty} \beta_k^{r+1} |z_k^0|^2$$

$$< \frac{\varepsilon}{3} + \frac{\varepsilon}{3} + \frac{\varepsilon}{3} = \varepsilon.$$

This proves property (iii).

Next since $\dfrac{4}{1 - e^{-2\beta_k t}} > 4$, $t > 0$, therefore we obtain from (i) and (iii) that

$$E(t) > 4 \sum_{k=1}^{\infty} \beta_k^{r+1} |z_k^0|^2, \quad t > 0. \tag{11}$$

Property 1 and (11) imply that the equation

$$E(t) = \rho_0^2 \tag{12}$$

has a root $t = \theta$ if and only if

$$\rho_0^2 > 4 \sum_{k=1}^{\infty} \beta_k^{r+1} |z_k^0|^2, \tag{13}$$

and this root is unique. Without loss of generality, we can assume that $\theta < T$ since T is sufficiently big number.

The following statement is a solution for the control problem (5).

Theorem 1 *Let inequality (13) be satisfied and $z^0, z^1 \in l_r^2$. Then the system (4) can be transferred from the initial position z^0 to the position z^1 for the time θ.*

Proof Define a control

$$w_k(t) = \begin{cases} -\left[z_k^0 - z_k^1 e^{-\beta_k \theta}\right] \phi_k(\theta) e^{-\beta_k t}, & 0 \le t \le \theta \\ 0, & t > \theta \end{cases}, \quad k = 1, 2, \ldots. \quad (14)$$

Show that this control is admissible. Using Eq. (12), control (14), and the obvious inequality $|x - y|^2 \le 2|x|^2 + 2|y|^2$, we proceed as follows:

$$
\sum_{k=1}^{\infty} \beta_k^r \int_0^\theta |w_k(s)|^2 ds = \sum_{k=1}^{\infty} \beta_k^r \int_0^\theta \left| -\left[z_k^0 - z_k^1 e^{-\beta_k \theta}\right] \phi_k(\theta) e^{-\beta_k s} \right|^2 ds
$$

$$
\le \sum_{k=1}^{\infty} \beta_k^r \left(2|z_k^0|^2 + 2|z_k^1|^2 e^{-2\beta_k \theta} \right) \phi_k^2(\theta) \int_0^\theta e^{-2\beta_k s} ds
$$

$$
= 2 \sum_{k=1}^{\infty} \beta_k^r \left(|z_k^0|^2 \phi_k(\theta) + |z_k^1|^2 \psi_k(\theta) \right)
$$

$$
= E(\theta) = \rho_0^2.
$$

Show that the system can be transferred from z^0 to z^1 for the time θ. Indeed,

$$
z_k(\theta) = e^{\beta_k \theta} \left(z_k^0 - \left[z_k^0 - z_k^1 e^{-\beta_k \theta}\right] \phi_k(\theta) \int_0^\theta e^{-2\beta_k s} ds \right)
$$

$$
= e^{\beta_k \theta} (z_k^1 e^{-\beta_k \theta}) = z_k^1.
$$

This completes the proof of Theorem 1.

4   Pursuit Differential Game Problem

In this section, we study pursuit differential game described by the Eq. (3). It is assumed that control resources of pursuer is greater than that of evader, that is $\rho > \sigma$.

We obtain from (3) that

$$
z_k(t) = e^{\beta_k t} \left(z_k^0 - \int_0^t u_k(s) e^{-\beta_k s} ds + \int_0^t v_k(s) e^{-\beta_k s} ds \right). \quad (15)
$$

In view of the previous section we can state that the equation

$$E(t) = 2\sum_{k=1}^{\infty} \beta_k^r \left(|z_k^0|^2 \phi_k(t) + |z_k^1|^2 \psi_k(t)\right) = (\rho - \sigma)^2 \qquad (16)$$

has a root $t = \theta_1$ if and only if

$$(\rho - \sigma)^2 > 4\sum_{k=1}^{\infty} \beta_k^{r+1} |z_k^0|^2, \qquad (17)$$

and this root is unique. We can assume, by selecting T if needed that $\theta_1 < T$.

Theorem 2 *Let (17) be satisfied and $z^0, z^1 \in l_r^2$. Then pursuit can be completed in the game (3) for the time θ_1.*

Proof Construct a strategy for the pursuer. Set

$$u_k(t, v) = \begin{cases} \left[z_k^0 - z_k^1 e^{-\beta_k \theta_1}\right] \phi_k(\theta_1) e^{-\beta_k s} + v_k(t), & 0 \le t \le \theta_1 \\ 0, & t > \theta_1 \end{cases}, \quad k = 1, 2, \ldots \ (18)$$

Show that strategy (18) is admissible. Applying the Minkowskii inequality, we have

$$\left(\sum_{k=1}^{\infty} \beta_k^r \int_0^{\theta_1} |u_k(s)|^2 \, ds\right)^{1/2} = \left(\sum_{k=1}^{\infty} \beta_k^r \int_0^{\theta_1} \left|\left(z_k^0 - z_k^1 e^{-\beta_k \theta_1}\right) \phi_k(\theta_1) e^{-\beta_k s} + v_k(s)\right|^2 ds\right)^{1/2}$$

$$\le \left(\sum_{k=1}^{\infty} \beta_k^r \int_0^{\theta_1} \left|\left(z_k^0 - z_k^1 e^{-\beta_k \theta_1}\right) \phi_k(\theta_1) e^{-\beta_k s}\right|^2 ds\right)^{1/2}$$

$$+ \left(\sum_{k=1}^{\infty} \beta_k^r \int_0^{\theta_1} |v_k(s)|^2 \, ds\right)^{1/2}$$

$$\le \left(\sum_{k=1}^{\infty} \beta_k^r |z_k^0 - z_k^1 e^{-\beta_k \theta_1}|^2 \phi_k^2(\theta_1) \int_0^{\theta_1} e^{-2\beta_k s} ds\right)^{1/2} + \sigma. \quad (19)$$

Using the obvious inequality $|x - y|^2 \le 2|x|^2 + 2|y|^2$ and Eq. (16), we obtain form (19) that

$$\left(\sum_{k=1}^{\infty} \beta_k^r \int_0^{\theta_1} |u_k(s)|^2 \, ds\right)^{1/2} \le \left(2\sum_{k=1}^{\infty} \beta_k^r \left(|z_k^0|^2 \phi_k(\theta_1) + |z_k^1|^2 \psi_k(\theta_1)\right)\right)^{1/2} + \sigma$$

$$= E^{1/2}(\theta_1) + \sigma$$

$$= \rho - \sigma + \sigma = \rho.$$

Thus the strategy (18) is admissible.

Next, we show that pursuit is completed at the time θ_1. Indeed, using (15) and strategy (18), we have

$$z_k(\theta_1) = e^{\beta_k \theta_1} \left(z_k^0 - \int_0^{\theta_1} \left(\left(z_k^0 - z_k^1 e^{-\beta_k \theta_1} \right) \phi_k(\theta_1) e^{-\beta_k s} + v_k(s) \right) e^{-\beta_k s} ds + \int_0^{\theta_1} v_k(s) e^{-\beta_k s} ds \right)$$

$$= e^{\beta_k \theta_1} \left(z_k^0 - \int_0^{\theta_1} \left(z_k^0 - z_k^1 e^{-\beta_k \theta_1} \right) \phi_k(\theta_1) e^{-2\beta_k s} ds \right)$$

$$= e^{\beta_k \theta_1} \left(z_k^0 - z_k^0 + z_k^1 e^{-\beta_k \theta_1} \right) = z_k^1.$$

The proof of the theorem is completed.

5 Conclusion

We have studied a pursuit differential game problem described by infinite system of 1st-order differential equations with negative coefficients in the space l_{r+1}^2. The control functions of players are subjected to integral constraints.

We have obtained a condition for which a control problem is solvable, also we have constructed a control that transfers the system from an initial state z^0 to the final state z^1 for a finite time.

We have obtained a condition of completion of pursuit in the differential game. Moreover, a pursuit strategy has been constructed.

Acknowledgements The present research was partially supported by the National Fundamental Research Grant Scheme FRGS of Malaysia, 01-01-17-1921FR.

References

1. Alias, I.A., Ibragimov, G.I., Rakhmanov, A.T.: Evasion differential game of infinitely many evaders from infinitely many pursuers in Hilbert space. Dyn. Games Appl. **7**, 347–359 (2017). https://doi.org/10.1007/s1323501601960
2. Avdonin, S.A., Ivanov, S.A.: Families of Exponentials: the Method of Moments in Controllability Problems for Distributed Parameter Systems. Cambridge University Press, Cambridge (1995)
3. Azamov, A.A., Ruziboyev, M.B.: The time-optimal problem for evolutionary partial differential equations. J. Appl. Maths Mekhs. **77**(2), 220–224 (2013)
4. Azamov, A.A., Samatov, B.T.: P-strategy. An Elementary Introduction to the Theory of Differential Games. National University Press, Tashkent (2000)
5. Belousov, A.A.: Method of resolving functions for differential games with integral constraints. Theory Opt. Solut. **9**, 10–16 (2010)
6. Butkovsky, A.G.: Theory of Optimal Control of Distributed Parameter Systems. Elsevier, New York (1964)
7. Chernous'ko, F.L.: Bounded controls in distributed-parameter systems. J. Appl. Maths Mekhs. **56**(5), 707–723 (1992)
8. Chikrii, A.A., Belousov, A.A.: On linear differential games with integral constraints. In: Proceedings of the Steklov Institute of Mathematics. 269 (Issue 1 Supplement), pp. 69–80 (2010)
9. Egorov, A.I.: Principles of the Control Theory. Nauka, Moscow (2004)

10. Friedman, A.: Differential Games. Wiley, New York (1971)
11. Guseinov, K.G., Neznakhin, A.A., Ushakov, V.N.: Approximate construction of attainability sets of control systems with integral constraints on the controls. J. Appl. Maths Mekhs. **63**(4), 557–567 (1999)
12. Huseyin, A., Huseyin, N.: Precompactness of the set of trajectories of the controllable system described by a nonlinear Volterra integral equation. Math. Model. Anal. **17**(5), 686–695 (2012)
13. Ibragimov, G.I.: A problem of optimal pursuit in systems with distributed parameters. J. Appl. Maths Mekhs. **66**(5), 719–724 (2002) (Prikladnaya Matematika i Mekhanika. **66**(5), 753–759 (2002))
14. Ibragimov, G.I.: The optimal pursuit problem reduced to an infinite system of differential equations. J. Appl. Maths Mekhs. **77**(5), 470–476 (2013). https://doi.org/10.1016/j.jappmathmech.2013.12.002
15. Ibragimov, G.I.: Optimal pursuit time for a differential game in the Hilbert space l_2. ScienceAsia **39S**, 25–30 (2013)
16. Ibragimov, G.I., Hasim, R.M.: Pursuit and evasion differential games in Hilbert space. Int. Game Theory Rev. **12**(03), 239–251 (2010)
17. Ibragimov, G.I., Ja'afaru, A.B.: On control problem described by infinite system of first order differential equations. Aust. J. Basic Appl. Sci. **5**(10), 736–742 (2011)
18. Ibragimov, G.I., Salleh, Y.: Simple motion evasion differential game of many pursuers and one evader with integral constraints on control functions of players. J. Appl. Maths. 2012 (Article ID 748096), 10 p (2012). https://doi.org/10.1155/2012/748096
19. Ibragimov, G.I., Azamov, A.A., Khakestari, M.: Solution of a linear pursuit-evasion game with integral constraints. ANZIAM J. **52**(E), E59–E75 (2011)
20. Ibragimov, G.I., Salimi, M., Amini, M.: Evasion from many pursuers in simple motion differential game with integral constraints. Euro. J. Oper. Res. **218**(2), 505–511 (2012)
21. Ibragimov, G.I., Satimov, N.Yu.: A multi player pursuit differential game on closed convex set with integral constraints. Abstr. Appl. Anal. 2012 (Article ID 460171), 12 p (2012). https://doi.org/10.1155/2012/460171
22. Ibragimov, G.I., Allahabi, F., Kuchkarov, A.: A pursuit problem in an infinite system of second-order differential equations. Ukr. Math. J. **65**(8), 1203–1216 (2014)
23. Ibragimov, G.I., Abd Rasid, N., Kuchkarov, A.Sh., Ismail. F.: Multi pursuer differential game of optimal approach with integral constraints on controls of players. Taiwanese J. Math. **19**(3), 963–976 (2015). https://doi.org/10.11650/tjm.19.2015.2288
24. Idham, A.A., Ibragimov, G.I., Kuchkarov, A.S., Akmal, S.: Differential game with many pursuers when controls are subjected to coordinate-wise integral constraints. Malays. J. Math. Sci. **10**(2), 195–207 (2016)
25. Isaacs, R.: Differential Games, a Mathematical Theory with Applications to Optimization, Control and Warfare. Wiley, New York (1965)
26. Konovalov, A.P.: Linear differential evasion games with lag and integral constraints. Cybernetics **4**, 41–45 (1987)
27. Krasovskii, N.N.: The Theory of Motion Control. Nauka, Moscow (1968)
28. Krasovskii, N.N., Subbotin, A.I.: Game-Theoretical Control Problems. Springer, New York (1988)
29. Kuchkarov, A.S.: On a differential game with integral and phase constraints. Autom. Remote Control **74**(1), 12–25 (2013)
30. Lewin, J.: Differential Games. Springer, London (1994)
31. Lokshin, M.D.: Differential games with integral constraints on disturbances. J. Appl. Maths Mekhs. **54**(3), 331–337 (1990)
32. Mamatov, M.S.: On the theory of pursuit games in distributed parameters systems. Autom. Control Comput. Sci. **43**(1), 1–8 (2008)
33. Mikhlin, S.G.: Linear Partial Differential Equations. Visshaya Shkola, Moscow (1977). (in Russian)
34. Nikolskii, M.S.: The direct method in linear differential games with integral constraints. Controlled systems, IM, IK, SO AN SSSR **2**, 49–59 (1969)

35. Okhezin, S.P.: Differential encounter-evasion game for parabolic system under integral constraints on the player's controls. Prikl Mat i Mekh. **41**(2), 202–209 (1977)
36. Petrosyan, L.A.: Differential Games of Pursuit. World Scientific, Singapore (1993)
37. Pontryagin, L.S.: Selected scientific works. **2** (1988)
38. Salimi, M., Ibragimov, G., Siegmund, S., Sharifi, S.: On a fixed duration pursuit differential game with geometric and integral constraints. Dyn. Games Appt. **6**, 409–425 (2016)
39. Samatov, B.T.: Problems of group pursuit with integral constraints on controls of the players I. Cybern. Syst. Anal. **49**(5), 756–767 (2013)
40. Satimov, N.Y., Tukhtasinov, M.: On some game problems distributed controlled system. J. Appl. Maths Mekhs. **69**, 885–890 (2005)
41. Satimov, N.Y., Tukhtasinov, M.: Game problems on a fixed interval in controlled first-order evolution equations. Maths Notes **80**(3–4), 578–589 (2006)
42. Satimov, N.Y., Rikhsiev, B.B., Khamdamov, A.A.: On a pursuit problem for n-person linear differential and discrete games with integral constraints. Math. USSR Sbornik **46**(4), 459–471 (1983)
43. Solomatin, A.M.: A game theoretic approach evasion problem for a linear system with integral constraints imposed on the player control. J. Appl. Maths Mekhs. **48**(4), 401–405 (1984)
44. Subbotin, A.I., Ushakov, V.N.: Alternative for an encounter-evasion differential game with integral constraints on the players' contols. J. Appl. Maths Mekhs. **39**(3), 367–375 (1975)
45. Tukhtasinov, M., Mamatov, M.S.: On pursuit problems in controlled distributed systems. Maths Notes **84**(1–2), 256–262 (2008)
46. Tukhtasinov, M., Mamatov, M.S.: On transfer problem in controlled system. Differ. Equ. **45**(3), 439–444 (2009)

Invariance Principles for Ergodic Systems with Slowly α-Mixing Inducing Base

Jianyu Chen and Kien Nguyen

Abstract We investigate a class of ergodic systems, which admit an inducing base with a slowly α-mixing generating partition. Under suitable moment condition on the first return time, we prove the almost sure invariance principle (ASIP) for adapted stationary processes. Our results apply to intermittent maps and billiards with flat points.

Keywords ASIP · Alpha-mixing · Inducing · Stationary process

1 Introduction

As a functional generalization of the central limit theorems, the almost sure invariance principle (ASIP) asserts the the partial sum of a random process can be well approximated by a Brownian motion with an almost sure error. There has been a great deal of work on the invariance principles in probability theory, such as [2, 8, 11, 22, 25, 27], etc., as well as in the context of dynamical systems, for instance, [1, 3–5, 9, 12, 14, 15, 20, 21, 23, 26, 28, 29], etc. Three major approaches are exploited in the proof of invariance principles: (1) the martingale approximation method (e.g. [5, 22]); (2) the inducing and Young towers (e.g. [20, 21]); (3) the spectral method for transfer operators (e.g. [12, 23]).

In the paper, we study the almost sure invariance principle (ASIP) for a class of ergodic dynamical systems with a slowly α-mixing inducing base. Our setting is rather abstract, and does not have any smooth structures. Also, we assume very

J. Chen (✉) · K. Nguyen
University of Massachusetts Amherst, Amherst, MA, USA
e-mail: jchen@math.umass.edu

K. Nguyen
e-mail: kien@math.umass.edu

© Springer Nature Switzerland AG 2018

A. Azamov et al. (eds.), *Differential Equations and Dynamical Systems*,
Springer Proceedings in Mathematics & Statistics 268,
https://doi.org/10.1007/978-3-030-01476-6_9

low regularity for the observable that generates the stationary process, that is, the observable is only integrable but could be unbounded. In this situation, we are able to prove the ASIP for stationary processes that are generated by any adapted observables. Although adapted observables might be a quite narrowed class of functions, they can provide good approximations for most regular observables.

This paper is organized as follows. In Sect. 2, we shall introduce Assumption (**H1**) on the inducing base and Assumption (**H2**) for the first return time, and state our main theorem. In Sect. 3, we deliver the proof of the ASIP in four subsections. In Sect. 4, we apply our main result to intermittent maps and billiards with flat points.

2   Statement of Results

Let $\mathcal{T}$ be an ergodic measure-preserving transformation on a standard probability space $(\mathcal{M}, \mathcal{B}, \mu)$. We choose a subset $M \subset \mathcal{M}$ of positive μ-measure, and denote the first return time to M by

$$R(x) = \inf\{n \geq 1 : \mathcal{T}^n(x) \in M\}, \quad \text{for any } x \in M.$$

Consider the induced base transformation $T : (M, \mathcal{B}_M, \nu) \circlearrowleft$, where

- $T(x) = \mathcal{T}^{R(x)}(x)$ for any $x \in M$;
- $\mathcal{B}_M := \{B \cap M : B \in \mathcal{B}\}$;
- ν is the conditional measure of μ on M, i.e., $\nu(\cdot) = \mu(\cdot \mid M)$.

By Poincaré recurrence and the ergodicity of $\mathcal{T}$, we have

$$M = \bigcup_{n=1}^{\infty}\{R = n\} \quad (\text{mod } \nu), \quad \text{and} \quad \mathcal{M} = \bigcup_{n=1}^{\infty}\bigcup_{k=0}^{n-1} \mathcal{T}^k\{R = n\} \quad (\text{mod } \mu).$$

Remark 1 The induced map T must be ergodic, since the original map $\mathcal{T}$ is ergodic. However, $\mathcal{T}$ may not be mixing, even if T is mixing.

We now impose the following assumptions.

(**H1**) T admits a generating partition ξ, i.e., $\mathcal{F}_0^{\infty} = \mathcal{B}_M$ (mod ν), where $\mathcal{F}_s^t :=$ $\sigma\left(T^{-s}\xi \vee \cdots \vee T^{-t}\xi\right)$ for any $0 \leq s \leq t \leq \infty$. Moreover, the family $\mathfrak{F} := \{\mathcal{F}_s^t\}_{0 \leq s \leq t \leq \infty}$ is α-mixing with polynomial rate $\mathcal{O}(n^{-\beta})$ for some $\beta > 2$, that is,

$$\alpha_{\mathfrak{F}}(n) = \sup_{t \geq 0} \sup_{A \in \mathcal{F}_0^t} \sup_{B \in \mathcal{F}_{t+n}^{\infty}} |\nu(A \cap B) - \nu(A)\nu(B)| = \mathcal{O}(n^{-\beta}). \tag{1}$$

(**H2**) $R \in L^p(M, \nu)$ for some $p > 2$ satisfying $\frac{1}{\beta} + \frac{1}{p} < \frac{1}{2}$, or equivalently,

$$\nu\{R > k\} = \mathcal{O}(k^{-p}). \tag{2}$$

Refining ξ if necessary, one may assume that $\{R = n\} \in \mathcal{F}_0^0$ for each $n \geq 1$. We then naturally lift the partition ξ to the partition $\widetilde{\xi}$ on $\mathcal{M}$, to be precise,

$$\widetilde{\xi} := \left\{ A \subset \mathcal{T}^k \{R = n\} : \mathcal{T}^{-k} A \in \xi, \, n \geq 1, \, 0 \leq k \leq n - 1 \right\}.$$

It is clear that $\widetilde{\xi}$ is a generating partition for $\mathcal{T}$. We denote $\widetilde{\mathcal{F}}_s^t := \sigma$ $\left(\mathcal{T}^{-s} \widetilde{\xi} \vee \cdots \vee \mathcal{T}^{-t} \widetilde{\xi} \right)$ for any $0 \leq s \leq t \leq \infty$.

A measurable function $f : \mathcal{M} \to \mathbb{R}$(or $f : M \to \mathbb{R}$) is said to be an *adapted* function if f is $\widetilde{\mathcal{F}}_s^t$-measurable (or $\mathcal{F}_s^t$-measurable) for some $0 \leq s \leq t < \infty$. In particular, the first return time R is adapted.

Our main result is the following.

Theorem 1 *Let $q > 2$ be such that $\frac{1}{\beta} + \frac{1}{p} + \frac{1}{q} < \frac{1}{2}$. Suppose that $f \in L^q(\mathcal{M}, \mu)$ with $\mathbb{E}_\mu(f) = 0$, and f is an adapted function on $\mathcal{M}$. Then the stationary process $\mathbf{X}_f := \{f \circ \mathcal{T}^n\}_{n \geq 0}$ satisfies an almost sure invariance principle (ASIP) as follows: for any $\lambda \in \left(\max \left\{ \frac{1}{4}, \frac{1}{\beta} + \frac{1}{p} + \frac{1}{q} \right\}, \frac{1}{2} \right)$, enlarging to a richer probability space $(\mathcal{M}', \mu')$ if necessary, there exists a standard Brownian motion $W(\cdot)$ such that*

$$\left| \sum_{k=0}^{n-1} f \circ \mathcal{T}^k - W(n\sigma^2) \right| = \mathcal{O}(n^\lambda), \quad \mu' - a.s. \tag{3}$$

where $\sigma = \sigma(f)$ is defined by (18) in Sect. 3.4.

It is obvious from (3) that $\sigma = \lim_{n \to \infty} \frac{1}{n} \mathbb{E}_\mu \left(\sum_{k=0}^{n-1} f \circ \mathcal{T}^k \right)^2$. We shall provide an alternative formula in (18) for σ from the induced system.

Remark 2 We could easily extend Theorem 1 in the invertible case, with the only modification on the families $\mathcal{F}_s^t$ and $\widetilde{\mathcal{F}}_s^t$ to be two sided, i.e., $-\infty \leq s \leq t \leq \infty$.

3 Proof of Theorem 1

3.1 The Induced Function $\widehat{f}$

For any measurable function $f : \mathcal{M} \to \mathbb{R}$, we define the induced function on M by

$$\widehat{f}(x) := \sum_{k=0}^{R(x)-1} f \circ \mathcal{T}^k(x), \quad x \in M.$$

Lemma 1 *Let $f : \mathcal{M} \to \mathbb{R}$ be a function that satisfies Theorem 1. Then*

(1) $\mathbb{E}_\nu(\widehat{f}) = 0$;

(2) $\widehat{f} \in L^r(M, \nu)$ *for any* $r \in \left(2, \frac{pq}{p+q}\right)$;

(3) For each $n \geq 0$, *the function* $\widehat{f} \circ T^n$ *is adapted on* M.

Proof (1) By Kac formula, i.e., $\int_M \widehat{f} d\mu = \int f d\mu$, and the fact that $\nu(\cdot) = \mu(\cdot|M)$, we have that $\mathbb{E}_\nu(\widehat{f}) = 0$ if $\mathbb{E}_\mu(f) = 0$.

(2) Note that $\widehat{f} = \sum_{k=0}^\infty f \circ \mathcal{T}^k \mathbb{1}_{\{R>k\}}$, then by Minkowski's inequality, Hölder inequality and $\mathcal{T}$-invariance of μ, we have

$$
\begin{aligned}
\|\widehat{f}\|_{L^r(\nu)} &\leq \sum_{k=0}^\infty \||f| \circ \mathcal{T}^k \mathbb{1}_{\{R>k\}}\|_{L^r(\nu)} \\
&= \mu(M)^{-\frac{1}{r}} \sum_{k=0}^\infty \left(\int |f|^r \circ \mathcal{T}^k \mathbb{1}_{\{R>k\}} d\mu\right)^{\frac{1}{r}} \\
&\leq \mu(M)^{-\frac{1}{r}} \sum_{k=0}^\infty \|f \circ \mathcal{T}^k\|_{L^q(\mu)} \left(\mu\{R > k\}\right)^{1/r - 1/q} \\
&= \mu(M)^{-\frac{1}{q}} \|f\|_{L^q(\mu)} \sum_{k=0}^\infty \left(\nu\{R > k\}\right)^{1/r - 1/q}.
\end{aligned}
$$

The last summation is finite due to Condition (2), i.e.,

$$
\sum_{k=0}^\infty \left(\mu\{R > k\}\right)^{1/r - 1/q} = 1 + \mathcal{O}\left(\sum_{k=1}^\infty \left(k^{-p}\right)^{1/r - 1/q}\right) < \infty,
$$

since $p(1/r - 1/q) > 1$. Therefore, $\|\widehat{f}\|_{L^r(\nu)} < \infty$ and thus $\widehat{f} \in L^r(M, \nu)$.

(3) Since f is adapted, there are $0 \leq s \leq t < \infty$ such that f is $\widehat{\mathcal{F}}_s^t$-measurable. It is easy to see that $\widehat{f}$ is $\mathcal{F}_s^t$-measurable. Moreover, we have that $\widehat{f} \circ T^n$ is $\mathcal{F}_{s+n}^{t+n}$-measurable for each $n \geq 0$, since $T^{-n}\mathcal{F}_s^t = \mathcal{F}_{s+n}^{t+n}$.

We shall first study the induced process $\mathbf{X}_{\widehat{f}} := \{\widehat{f} \circ T^n\}_{n \geq 1}$ on (M, ν).

3.2 ASIP for the Induced Process $\mathbf{X}_{\widehat{f}}$

In this subsection, we establish an ASIP for the induced process $\mathbf{X}_{\widehat{f}} = \{\widehat{f} \circ T^n\}_{n \geq 1}$. We first recall the following special case of an ASIP result by Shao and Lu [25].

Definition 1 Given a random process $\mathbf{X} = \{X_n\}_{n \geq 0}$ on (M, ν), we denote

$$
\mathcal{G}_m^n(\mathbf{X}) := \sigma\{X_m, X_{m+1}, \ldots, X_n\}
$$

for any $0 \leq m \leq n \leq \infty$. The α-mixing coefficient of the process is defined by

$$\alpha_{\mathbf{X}}(n) = \sup_{k \geq 0} \sup_{A \in \mathcal{G}_0^k(\mathbf{X})} \sup_{B \in \mathcal{G}_{k+n}^\infty(\mathbf{X})} |\nu(A \cap B) - \nu(A)\nu(B)|.$$

Proposition 1 *Let $\delta \in (0, 2]$ and $r \in (2 + \delta, \infty]$. If $\mathbf{X} = \{X_n\}_{n \geq 0}$ is a zero-mean random process such that*

(i) $\sup_{n \geq 0} \|X_n\|_{L^r} < \infty$;

(ii) $\sum_{n=1}^\infty \alpha_{\mathbf{X}}(n)^{\frac{1}{2+\delta} - \frac{1}{r}} < \infty$;

(iii) $\liminf_{n \to \infty} \dfrac{a_n}{n} > 0$, where $a_n := \mathbb{E}_\nu(\sum_{k=0}^{n-1} X_k)^2$,

then for any $\epsilon > 0$, enlarging to a richer probability space (M', ν') if necessary, there exists a standard Brownian motion $W(\cdot)$ such that

$$\left| \sum_{k=0}^{n-1} X_k - W(a_n) \right| = \mathcal{O}\left(a_n^{\frac{1}{2+\delta} + \epsilon} \right), \quad \nu' - a.s.$$

We now directly apply Proposition 1 to adapted stationary processes on (M, ν).

Lemma 2 *Let $r > 2$ be such that $\frac{1}{\beta} + \frac{1}{r} < \frac{1}{2}$. Suppose that $g \in L^r(M, \nu)$ with $\mathbb{E}_\nu(g) = 0$, and g is an adapted function on M. Then the stationary process $\mathbf{X}_g = \{g \circ T^n\}_{n \geq 0}$ satisfies an ASIP as follows: for any $\lambda \in \left(\max\left\{ \frac{1}{4}, \frac{1}{\beta} + \frac{1}{r} \right\}, \frac{1}{2} \right)$, enlarging to a richer probability space (M', ν') if necessary, there exists a standard Brownian motion $W(\cdot)$ such that*

$$\left| \sum_{k=0}^{n-1} g \circ T^k - W\left(n\sigma_g^2\right) \right| = \mathcal{O}\left(n^\lambda\right), \quad \nu' - a.s. \tag{4}$$

where σ_g^2 is given by

$$\sigma_g^2 := \sum_{n=-\infty}^\infty \mathbb{E}_\nu(g \cdot g \circ T^n) = \sum_{n=-\infty}^\infty \int g \cdot g \circ T^n \, d\nu. \tag{5}$$

Proof In the degenerate case when $\sigma_g = 0$, it is well known that g is a coboundary, i.e., there exists a measurable function $h : M \to \mathbb{R}$ such that $g = h - h \circ T$ (see e.g. [17], Theorem 18.2.2), and thus (8) is automatic.

We now consider the non-degenerate case when $\sigma_g > 0$, and check conditions in Proposition 1 for the stationary process $\mathbf{X}_g := \{g \circ T^n\}_{n \geq 0}$ as follows.

As $\lambda \in \left(\max\left\{ \frac{1}{4}, \frac{1}{\beta} + \frac{1}{r} \right\}, \frac{1}{2} \right)$, we pick a sufficiently small $\delta \in \left(0, \frac{1}{\lambda} - 2\right)$ such that $\frac{1}{r} < \frac{1}{2+\delta} - \frac{1}{\beta}$. By T-invariance of ν, we have $\mathbb{E}_\nu(g \circ T^n) = \mathbb{E}_\nu(g) = 0$ for any $n \geq 0$, that is, the process is of zero mean. Also, $\|g \circ T^n\|_{L^r(\nu)} = \|g\|_{L^r(\nu)}$, and thus Condition (i) in Proposition 1 holds.

For Condition (ii), we recall that $\mathcal{G}_m^n(\mathbf{X}_g)$ is the σ-algebra generated by $g \circ T^m, \ldots, g \circ T^n$, where $0 \leq m \leq n \leq \infty$. Since g is an adapted function, there are some $0 \leq s \leq t < \infty$ such that g is $\mathcal{F}_s^t$-measurable. Therefore, $g \circ T^n$ is $\mathcal{F}_{s+n}^{t+n}$-measurable, and hence $\mathcal{G}_m^n(\mathbf{X}_g) \subset \mathcal{F}_{s+m}^{t+n}$. Hence by (1),

$$\alpha_{\mathbf{X}_g}(n) \leq \alpha_{\mathfrak{F}}(n + s - t) = \mathcal{O}\left((n+s-t)^{-\beta}\right) = \mathcal{O}\left(n^{-\beta}\right),$$

as $n \to \infty$, which immediately implies Condition (ii) since $\beta\left(\frac{1}{2+\delta} - \frac{1}{r}\right) > 1$.

By the covariance inequality in Lemma 7.2.1 in [22], we have

$$\left|\mathbb{E}_v(g \cdot g \circ T^n)\right| \leq 10\alpha_{\mathbf{X}_g}(n)^{1-\frac{2}{r}}\|g\|_{L^r(v)}\|g \circ T^n\|_{L^r(v)}$$
$$\leq 10\|g\|_{L^r(v)}^2\, \mathcal{O}\left(n^{-\beta(1-\frac{2}{r})}\right) =: \mathcal{O}\left(n^{-\beta_1}\right),$$

where we set $\beta_1 := \beta(1 - \frac{2}{r}) > 2$. Hence the series in (5) absolutely converges. We now check Condition (iii).

$$a_n = \mathbb{E}_v\left(\sum_{k=0}^{n-1} g \circ T^k\right)^2 = n\mathbb{E}_v(g)^2 + 2\sum_{k=1}^{n-1}(n-k)\mathbb{E}_v(g \cdot g \circ T^k)$$

$$= n\sigma_g^2 - n\sum_{|k| \geq n}\mathbb{E}_v(g \cdot g \circ T^k) - 2\sum_{k=1}^{n-1}k\mathbb{E}_v(g \cdot g \circ T^k)$$

$$= n\sigma_g^2 + \mathcal{O}\left(n\sum_{|k| \geq n}k^{-\beta_1}\right) + \mathcal{O}\left(\sum_{k=1}^{n-1}k^{1-\beta_1}\right)$$

$$= n\sigma_g^2 + \mathcal{O}(1),$$

Therefore, $\lim_{n \to \infty} \frac{a_n}{n} = \sigma_g^2 > 0$.

By Proposition 1, for any $\epsilon \in (0, \lambda - \frac{1}{2+\delta})$, enlarging to a richer probability space (M', v') if necessary, there exists a standard Brownian motion $W(\cdot)$ such that

$$\left|\sum_{k=0}^{n-1} g \circ T^k - W(a_n)\right| = \mathcal{O}\left(n^{\frac{1}{2+\delta}+\epsilon}\right) = \mathcal{O}(n^\lambda), \quad v' - a.s. \tag{6}$$

We recall the following property of standard Brownian motions: for any $s \geq 0$ and $t > 0$, the increment $W(s + t) - W(s)$ has the same distribution as $Z(t)$, where $Z(t)$ is normally distributed with mean 0 and variance t. Also, it is well known that $\mathbb{E}|Z(t)|^{2\ell} = t^\ell(2\ell - 1)!!$ for any $\ell \in \mathbb{N}$, where the double factorial is defined by $(2\ell - 1)!! = \prod_{k=1}^{\ell}(2k - 1)$. In particular, $\mathbb{E}|Z(t)|^4 = 3t^2$. See e.g. [10] for details.

Now we compare $W(a_n)$ and $W\left(n\sigma_g^2\right)$ as follows. Since $a_n = n\sigma_g^2 + \mathcal{O}(1)$, by Markov's inequality,

$$\sum_{n=1}^{\infty} v' \left\{ \left| W(a_n) - W\left(n\sigma_g^2\right) \right| \geq n^{\lambda} \right\} \leq \sum_{n=1}^{\infty} \frac{\mathbb{E}_{v'} \left| Z(|a_n - n\sigma_g^2|) \right|^4}{n^{4\lambda}}$$

$$= \sum_{n=1}^{\infty} n^{-4\lambda} \cdot 3 \left| a_n - n\sigma_g^2 \right|^2$$

$$= \mathcal{O}\left(\sum_{n=1}^{\infty} n^{-4\lambda} \right) < \infty,$$

as $\lambda > \frac{1}{4}$. Then by Borel–Cantelli Lemma,

$$\left| W(a_n) - W\left(n\sigma_g^2\right) \right| = \mathcal{O}(n^{\lambda}), \quad v' - a.s. \tag{7}$$

Therefore, (4) immediately follows from (6) and (7).

Applying Lemma 2 to the induced processes, we obtain

Lemma 3 *The induced process $\mathbf{X}_{\widehat{f}} = \{\widehat{f} \circ T^n\}_{n\geq 0}$ satisfies an ASIP as follows: for any $\lambda \in \left(\max\left\{ \frac{1}{4}, \frac{1}{\beta} + \frac{1}{p} + \frac{1}{q} \right\}, \frac{1}{2} \right)$, enlarging to a richer probability space (M', v') if necessary, there exists a standard Brownian motion $W(\cdot)$ such that*

$$\left| \sum_{k=0}^{n-1} \widehat{f} \circ T^k - W\left(n\sigma_{\widehat{f}}^2\right) \right| = \mathcal{O}(n^{\lambda}), \quad v' - a.s. \tag{8}$$

where $\sigma_{\widehat{f}}^2$ is given by (5).

Proof Recall that $\lambda \in \left(\max\left\{ \frac{1}{4}, \frac{1}{\beta} + \frac{1}{p} + \frac{1}{q} \right\}, \frac{1}{2} \right)$. Pick a sufficiently small $\delta \in \left(0, \frac{1}{\lambda} - 2 \right)$, and choose some $r > 2$ such that

$$\frac{1}{p} + \frac{1}{q} < \frac{1}{r} < \lambda - \frac{1}{\beta}. \tag{9}$$

By Lemma 1, $\widehat{f} \in L^r(M, r)$ and $\mathbb{E}_v(\widehat{f}) = 0$, and $\widehat{f}$ is an adapted function on M. Then (8) holds by Lemma 2.

3.3 Comparison Between $\mathbf{X}_f$ and $\mathbf{X}_{\widehat{f}}$

We now regard v as a probability measure on $\mathcal{M}$, although it is not $\mathcal{T}$-invariant. Note that v-a.s. $x \in \mathcal{M}$ belongs to the induced space M. In this subsection, we shall show that the induced process $\mathbf{X}_{\widehat{f}} = \{\widehat{f} \circ T^n\}_{n\geq 1}$ on (M, v) is comparable to the original process $\mathbf{X}_f = \{f \circ \mathcal{T}^n\}_{n\geq 1}$ on $(\mathcal{M}, v)$.

For any point $x \in M$, or equivalently, for ν-a.s. $x \in \mathcal{M}$, we define the following time functions: for any $n \geq 1$, there is a unique integer $\widehat{n} = \widehat{n}(x, n)$ such that

$$\widehat{n} = \widehat{n}(x, n) := \max \left\{ m \geq 1 : \sum_{k=0}^{m-1} R \circ T^k(x) \leq n \right\}. \tag{10}$$

We set $\widehat{n} = 0$ if the above set is empty. Also, we let

$$\widetilde{n} = \widetilde{n}(x, n) := n - \sum_{k=0}^{\widehat{n}-1} R \circ T^k(x). \tag{11}$$

Lemma 4 *For any $\epsilon > 0$, we have*

$$|\widehat{n} - n\mu(M)| = \mathcal{O}\left(n^{\frac{1}{2}+\varepsilon}\right), \quad \nu - a.s. \tag{12}$$

Proof We first apply Lemma 2 to the stationary process

$$\mathbf{X}_R := \left\{ R \circ T^m - \mathbb{E}_\nu(R) \right\}_{m \geq 0} = \left\{ (R - \mathbb{E}_\nu(R)) \circ T^m \right\}_{m \geq 0}$$

on the probability space (M, ν). Indeed, $R - \mathbb{E}_\nu(R) \in L^p(\nu)$ and it is of zero mean. Furthermore, R is $\mathcal{F}_0^0$-measurable, and so is $R - \mathbb{E}_\nu(R)$. Hence by Lemma 2, enlarging to a richer probability space (M', ν') if necessary, there exists a standard Brownian motion $W_1(\cdot)$ such that

$$\left| \sum_{k=0}^{m-1} R \circ T^k - m\mathbb{E}_\nu(R) - W_1\left(m \, \sigma_{R-\mathbb{E}_\nu(R)}^2\right) \right| = \mathcal{O}\left(m^{\frac{1}{2}}\right), \quad \nu' - a.s. \tag{13}$$

By Kac formula, we have $\mathbb{E}_\nu(R) = \frac{1}{\mu(M)}$. It is well known (or use Borel–Cantelli Lemma) that for any $\varepsilon > 0$, $W_1\left(m \, \sigma_{R-\mathbb{E}_\nu(R)}^2\right) = \mathcal{O}\left(m^{\frac{1}{2}+\varepsilon}\right)$, ν'-a.s.. Hence (13) implies that

$$\sum_{k=0}^{m-1} R \circ T^k = \frac{m}{\mu(M)} + \mathcal{O}\left(m^{\frac{1}{2}+\varepsilon}\right), \quad \nu - a.s. \tag{14}$$

By the definitions in (10) and (11), we have

$$|\widetilde{n}| = \left| n - \sum_{k=0}^{\widehat{n}-1} R \circ T^k \right| \leq R \circ T^{\widehat{n}} = \mathcal{O}\left(\widehat{n}^{\frac{1}{p}+\varepsilon}\right) = \mathcal{O}\left(\widehat{n}^{\frac{1}{2}}\right), \quad \nu - a.s. \tag{15}$$

where we use that $R \in L^p(\nu)$ and $p > 2$. Hence by (14) and (15),

$$n = \frac{\widehat{n}}{\mu(M)} + \mathcal{O}\left(\widehat{n}^{\frac{1}{2}+\varepsilon}\right),$$

for ν-a.s. $x \in M$. In particular, it follows that $\widehat{n} \to \infty$ a.s. if and only if $n \to \infty$, and $\widehat{n} = \mathcal{O}(n)$. Therefore,

$$n = \frac{\widehat{n}}{\mu(M)} + \mathcal{O}\left(n^{\frac{1}{2}+\varepsilon}\right), \quad \nu - a.s.$$

from which (12) holds.

To compare the partial sums of $\mathbf{X}_f$ and $\mathbf{X}_{\widehat{f}}$, we consider

$$\Delta_n(x) := \sum_{k=0}^{n-1} f \circ \mathcal{T}^k(x) - \sum_{j=0}^{\widehat{n}-1} \widehat{f} \circ T^j(x) = \sum_{k=0}^{\widetilde{n}-1} f \circ \mathcal{T}^k(T^{\widehat{n}}(x)). \tag{16}$$

for ν-a.s. $x \in M$.

Set $h = |f|$, and let $\widehat{h}$ be its induced function on M. Let λ be given by Theorem 1. We choose r as in (9) and pick a sufficiently small $\varepsilon > 0$ such that $\frac{1}{r} + \varepsilon < \lambda$. Since $h = |f| \in L^q(\mathcal{M}, \mu)$, by the same argument in the proof of Lemma 1 (2), $\widehat{h} \in L^r(M, \nu)$. By Lemma 4 and the expression in (16), we get

$$|\Delta_n| \leq \widehat{h} \circ T^{\widehat{n}} = \mathcal{O}\left(\widehat{n}^{\frac{1}{r}+\varepsilon}\right) = \mathcal{O}\left(n^\lambda\right), \quad \nu - a.s. \tag{17}$$

3.4 ASIP for the Original Process

We set

$$\sigma = \sigma(f) := \sigma_{\widehat{f}} \sqrt{\mu(M)}. \tag{18}$$

where $\sigma_{\widehat{f}}$ is given by (5) (in which we let $g = \widehat{f}$).

Lemma 5 *For any $\varepsilon > 0$ and any standard Brownian motion $W(\cdot)$ on $(\mathcal{M}, \mu)$,*

$$\left| W\left(n\sigma^2\right) - W\left(\widehat{n}\sigma_{\widehat{f}}^2\right) \right| = \mathcal{O}(n^{\frac{1}{4}+\varepsilon}), \quad a.s., \tag{19}$$

Proof Pick a positive integer $\ell > 1/\varepsilon$. By the basic property of standard Brownian motions, as well as Markov's inequality, Lemma 4 and (18),

$$\sum_{n=1}^{\infty} \mu \left\{ \left| W\left(n\sigma^2\right) - W\left(\widehat{n}\sigma_{\widehat{f}}^2\right) \right| \geq n^{\frac{1}{4}+\varepsilon} \right\} \leq \sum_{n=1}^{\infty} \frac{\mathbb{E}_\mu \left| Z\left(\left| n\sigma^2 - \widehat{n}\sigma_{\widehat{f}}^2 \right| \right) \right|^{2\ell}}{n^{2\ell(\frac{1}{4}+\varepsilon)}}$$

$$= \sum_{n=1}^{\infty} n^{-2\ell(\frac{1}{4}+\varepsilon)} \cdot (2\ell-1)!! \left| n\sigma^2 - \widehat{n}\sigma_{\widehat{f}}^2 \right|^{\ell}$$

$$= \mathcal{O}\left(\sum_{n=1}^{\infty} n^{-\ell\varepsilon} \right) < \infty.$$

Here again $Z(t)$ denotes the normal distribution with mean 0 and variance t. Then (19) follows from the Borel–Cantelli Lemma.

Let λ be given by Theorem 1. Again we regard ν as a probability measure on $\mathcal{M}$, and we show that the original process $\{f \circ \mathcal{T}^k\}_{n\geq 0}$ satisfies an ASIP with rate $\mathcal{O}(n^\lambda)$ with respect to the measure ν.

Note that the almost sure bound for $|\Delta_n|$ in (17) also holds with respect to ν since ν is absolutely continuous with respect to μ. Then by Lemmas 3–5, enlarging $(\mathcal{M}, \nu)$ to a richer probability space $(\mathcal{M}', \nu')$ if necessary, there is a standard Brownian motion $W(\cdot)$ such that

$$\left| \sum_{k=0}^{n-1} f \circ \mathcal{T}^k - W\left(n\sigma^2\right) \right|$$

$$\leq \left| \sum_{k=0}^{n-1} f \circ \mathcal{T}^k - \sum_{j=0}^{\widehat{n}-1} \widehat{f} \circ T^j \right| + \left| \sum_{j=0}^{\widehat{n}-1} \widehat{f} \circ T^j - W\left(\widehat{n}\sigma_{\widehat{f}}^2\right) \right| + \left| W\left(\widehat{n}\sigma_{\widehat{f}}^2\right) - W\left(n\sigma^2\right) \right|$$

$$= \mathcal{O}(n^\lambda) + \mathcal{O}(\widehat{n}^\lambda) + \mathcal{O}(n^{\frac{1}{4}+\varepsilon}) = \mathcal{O}(n^\lambda), \quad \nu' - a.s.$$

Finally, we need to show the ASIP for the original process $\{f \circ \mathcal{T}^k\}_{n\geq 0}$ with respect to the original measure μ, as the Brownian motion $W(\cdot)$ is not defined in a richer space of $(\mathcal{M}, \mu)$. Nevertheless, this issue is recently solved by Korepanov [18] and Gouëzel [13]. Here we quote and state Corollary 1.3 in [13] for the our ergodic system $\mathcal{T} : (\mathcal{M}, \mu) \to (\mathcal{M}, \mu)$ with respect to the two measures ν and μ.

Proposition 2 *If the ASIP holds for the process $\{f \circ \mathcal{T}^k\}_{n\geq 0}$ with rate $\mathcal{O}(n^\lambda)$ with respect to ν, and $f \circ \mathcal{T}^n = \mathcal{O}(n^\lambda)$ a.s., with respect to both μ and ν, then the ASIP holds for $\{f \circ \mathcal{T}^k\}_{n\geq 0}$ with the same rate $\mathcal{O}(n^\lambda)$ with respect to μ.*

Applying this proposition, we finish the proof of Theorem 1 by confirming $f \circ \mathcal{T}^n = \mathcal{O}(n^\lambda)$. This is due to the fact that $f \in L^q$ and that $\lambda > \frac{1}{q}$.

4 Applications

4.1 Intermittent Maps

A classical example of one-dimensional intermittent maps is provided by the Manneville–Pomeau map $\mathcal{T}_\alpha : [0, 1] \to [0, 1]$ defined by

$$\mathcal{T}_\alpha(x) = x + x^{1+\alpha} \quad (\text{mod } 1),$$

for any $\alpha \in (0, 1)$. It was shown in [16, 19, 24, 29] that bounded Lipschitz observables has the correlation decay in rate $\mathcal{O}\left(n^{1-\frac{1}{\alpha}}\right)$, and satisfies the central limit theorem for $\alpha \in (0, 1/2)$. In [23], Pollicott and Sharp proved the weak invariance principle for $\alpha \in (0, 1/3)$.

We consider the case when $\alpha \in (0, \frac{1}{2})$. We obtain the induced map T_α on $M = [c, 1]$, where $c \in (0, 1)$ is such that $\mathcal{T}_\alpha(c) = 0$. It is well known that the first return time $R \in L^{1/\alpha}$, and the natural partition $\xi := \{[R = n]\}_{n \geq 1}$ is α-mixing with exponential rate. An observable f is adapted if there are $0 \leq s \leq t < \infty$ such that f is constant on each element of $T_\alpha^{-s} \vee \cdots \vee T_\alpha^{-t}$. By Theorem 1, the ASIP holds for any L^q adapted function with $q > \frac{\alpha}{1-2\alpha}$.

Remark 3 Of course, here we do not improve results in [23], since we only deal with adapted functions. Nevertheless, we do include some important functions, such as the first return time R itself, and thus our theorem provides an advanced result on the return time distribution.

4.2 Billiards with Flat Points

For the basics of chaotic billiards, we refer the reader to [7].

Chernov and Zhang [6] introduced a family of semi-dispersing billiards, for which the decay of correlations for the collision map $\mathcal{T}$ is of order $\mathcal{O}(n^{-a})$ for any $a \in (1, \infty)$. By carefully choosing an inducing domain M, they obtained a generating partition ξ of M given by the first return time $R \in L^{1+a}$. Also, the two-sided σ-filtration exhibits α-mixing with exponential rate. By Remark 2, our main theorem implies that the ASIP holds for any L^q adapted function with $q > 2\frac{a+1}{a-1}$.

References

1. Balint, P., Chernov, N., Dolgopyat, D.: Limit theorems for dispersing billiards with cusps. Comm. Math. Phys. **308**, 479–510 (2011)
2. Berkes, I., Philipp, W.: Approximation theorems for independent and weakly dependent random vectors. Ann. Probab. **7**(1), 29–54 (1979)

3. Bunimovich, L.A., Sinai, Y.G., Chernov, N.I.: Statistical properties of two-dimensional hyperbolic billiards. Russian Math. Surveys. **46**, 47–106 (1991)
4. Chernov, N.I.: Limit theorems and Markov approximations for chaotic dynamical systems. Probab. Theory Related Fields **101**(3), 321–362 (1995)
5. Chernov, N.I.: Advanced statistical properties of dispersing billiards. J. Statist. Phys. **122**, 1061–1094 (2006)
6. Chernov, N., Zhang, H.-K.: A family of chaotic billiards with variable mixing rates. Stochast. Dynam. **5**, 535–553 (2005)
7. Chernov, N.I., Markarian, R.: Chaotic Billiards, Math. Surveys Monographs, vol. 127, AMS, Providence (2006)
8. Cuny, C., Merlevède, F.: Strong invariance principles with rate for "reverse" martingale differences and applications. J. Theoret. Probab. **28**(1), 137–183 (2015)
9. Demers, M., Zhang, H.-K.: Spectral analysis of hyperbolic systems with singularities. Nonlinearity **27**, 379–433 (2014)
10. Durrett', R.: Probability: theory and examples, Fourth edn. Cambridge Series in Statistical and Probabilistic Mathematics, vol. 31, x+428 pp. Cambridge University Press, Cambridge (2010)
11. Eberlein, E.: On strong invariance principles under dependence assumptions. Ann. Probab. **14**(1), 260–270 (1986)
12. Gouëzel, S.: Almost sure invariance principle for dynamical systems by spectral methods. Ann. Probab. **38**(4), 1639–1671 (2010)
13. Gouëzel, S.: Variations around Eagleson's Theorem on mixing limit theorems for dynamical systems (Preprint, 2018)
14. Haydn, N., Vaienti, S.: Fluctuations of the metric entropy for mixing measures. Stoch. Dyn. **4**(4), 595–627 (2004)
15. Haydn, N., Nicol, M., Török, A., Vaienti, S.: Almost sure invariance principle for sequential and non-stationary dynamical systems. Trans. Amer. Math. Soc. **369**(8), 5293–5316 (2017)
16. Hu, H.: Decay of correlations for piecewise smooth maps with indifferent fixed points. Ergod. Theory Dyn. Syst. **24**, 495–524 (2004)
17. Ibragimov, I.A., Linnik, Y.V.: Independent and stationary sequences of random variables. Wolters-Noordhoff, Gröningen (1971)
18. Korepanov, A.: Equidistribution for nonuniformly expanding dynamical systems, and application to the almost sure invariance principle. Comm. Math. Phys. **359**(3), 1123–1138 (2018)
19. Liverani, C., Saussol, B., Vaienti, S.: A probabilistic approach to intermittency. Ergod. Theory Dyn. Syst. **19**(3), 671–685 (1999)
20. Melbourne, I., Nicol, M.: Almost sure invariance principle for nonuniformly hyperbolic systems. Commun. Math. Phys. **260**, 131–146 (2005)
21. Melbourne, I., Nicol, M.: A vector-valued almost sure invariance principle for hyperbolic dynamical systems. Ann. Probab. **37**(2), 478–505 (2009)
22. Philipp, W., Stout, W.: Almost sure invariance principles for partial sums of weakly dependent random variables, Memoir. Amer. Math. Soc. **161** (1975)
23. Pollicott, M., Sharp, R.: Invariance principle for interval maps with and indifferent fixed point. Comm. Math. Phys. **229**, 337–346 (2002)
24. Sarig, O.: Subexponential decay of correlations. Invent. Math. **150**, 629–653 (2002)
25. Shao, Q.M., Lu, C.R.: Strong approximations for partial sums of weakly dependent random variables. Sci. Sinica Ser. A **30**(6), 575–587 (1987)
26. Szász, D., Varjú, T.: Local limit theorem for Lorentz process and its recurrence in the plane. Ergod. Theory Dyn. Syst. **24**, 257–278 (2004)
27. Wu, W.B.: Strong invariance principles for dependent random variables. Ann. Probab. **35**(6), 2294–2320 (2007)
28. Young, L.S.: Statistical properties of systems with some hyperbolicity including certain billiards. Ann. Math. **147**, 585–650 (1998)
29. Young, L.S.: Recurrence times and rates of mixing. Israel J. Math. **110**, 153–188 (1999)

Central Limit Theorem for Billiards with Flat Points

Kien Nguyen and Hong-Kun Zhang

Abstract In this paper, we constructed stationary martingale difference approximations to certain processes generated by billiards with flat points, using the filtration generated by the first return time function. This leads to the central limit theorem for observables adapted to the filtration. Moreover, we also are able to obtain an explicit formula for the diffusion constant for this class of observables.

Keywords Billiards · Flat points · Limit theorems · Decay of correlations
Martingales

1 Introduction to the Main Result

Billiards are natural models to many different physical problems, especially in classical and statistical mechanics. They have a wide range of properties depending on the shape of the tables. Sinai introduced in 1970 the so-called Sinai (or dispersing) billiards where the boundary of the table is smooth and concave with positive curvature. These billiards are strongly chaotic: they are ergodic, mixing and have exponential decay of correlations. The central limit theorem is known to be true for these systems, see [2]. Since then, the central limit theorem and other limit theorems have been proved for various billiards, including ones with slow mixing rate. In many cases, the observables considered in those examples are Hölder continuous and the diffusion constant is given as an infinite series by the Green–Kubo formula.

In their paper [4], Chernov and Zhang introduced a family of dispersing billiard models. They were able to prove that the correlations for the collision map decay as $\mathcal{O}(1/n^a)$ for any constant $a \in (1, \infty)$, by introducing an induced system together with

K. Nguyen (✉) · H.-K. Zhang
University of Massachusetts Amherst, Amherst, MA, USA
e-mail: kien@math.umass.edu

H.-K. Zhang
e-mail: hongkun@math.umass.edu

© Springer Nature Switzerland AG 2018
A. Azamov et al. (eds.), *Differential Equations and Dynamical Systems*,
Springer Proceedings in Mathematics & Statistics 268,
https://doi.org/10.1007/978-3-030-01476-6_10

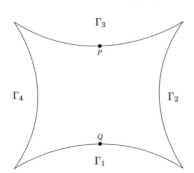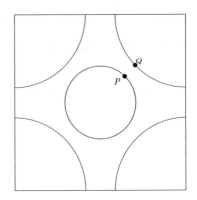

Fig. 1 Two types of dispersing billiards with one pair of flat points P and Q

a first return time function. Instead of using the traditional methods, we constructed a filtration generated by the first return time function. Then we are able to construct a stationary martingale difference sequence to approximate the process adapted to this filtration. With this new tool, we are going to the central limit theorem for this billiard family for a class of piecewise Hölder continuous functions. One achievement of our results is that we are able to represent the diffusion constants in an explicit and simple formula, comparing to the infinite series using the Green Kubo formula. Before we proceed to the main result, let us briefly recall some basic notions; more detailed exposition can be found in, for example, [3].

The billiard table $\mathcal{D}$ considered in [4] has a pair of boundary components bounded by the curves $y = |x|^\beta + 1$, $y = -(|x|^\beta + 1)$ and some strictly inward convex curves with nowhere vanishing curvature and no cusps. We denote P and Q be the only two points with zero curvature (also called flat points). It was assume that there is a periodic-2 trajectory running between the two flat points P and Q. A point mass moves inside the table and bounces off its boundary $\partial\mathcal{D}$ elastically, see Fig. 1.

Let $\mathcal{M}$ be the collision space of the billiard dynamics on $\mathcal{D}$. We parameterize $\partial\mathcal{D}$ by arclength in the clockwise direction and thus each collision is determined by its position r on $\partial\mathcal{D}$ and its angle of reflection $-\pi/2 \leq \varphi \leq \pi/2$ (that formed with the inward normal vector). They are natural coordinates $\mathcal{M}$ and we can write $\mathcal{M} = [0, |\partial\mathcal{D}|] \times [-\pi/2, \pi/2]$, where $|\partial\mathcal{D}|$ is the length of $\partial\mathcal{D}$. We denote S_0 as the boundary of $\mathcal{M}$, and S_1 as the singular set for F. The collision map $\mathcal{F} : \mathcal{M} \to \mathcal{M}$ preserves a smooth probability measure μ on $\mathcal{M}$ defined by:

$$d\mu = \frac{1}{2|\partial\mathcal{D}|} \cos(\varphi) dr d\varphi. \tag{1}$$

Let $f, g \in L^2(\mathcal{M}, \mu)$ be two piecewise Hölder continuous with singularities coincide with those of $\mathcal{F}^k$ for some k. The correlations of f and g are defined by:

$$C_n(f, g, \mathcal{F}, \mu) = \int_{\mathcal{M}} (f \circ \mathcal{F}^n) \cdot g \, d\mu - \int_{\mathcal{M}} f \, d\mu \int_{\mathcal{M}} g \, d\mu. \tag{2}$$

Chernov and Zhang proved in [4] that these correlations decay polynomially, that is:

$$|C_n(f, g, \mathcal{F}, \mu)| \leq C \frac{(\ln n)^{a+1}}{n^a}, \tag{3}$$

where $a = \frac{\beta+2}{\beta-2}$ and C is some fixed constant.

For systems with slow rates of decay of correlations like this, it is typical to study the dynamics on a subset of the phase space such that the induced system has exponential decay of correlations, then extend the results to the original space.

Let $M \subset \mathcal{M}$ be a subset of $\mathcal{M}$ obtained by removing the collisions that happen in an arbitrarily small neighbourhood of the flat points. The first return time function $R : M \to \mathbf{N}$ is defined almost everywhere by:

$$R(z) = \inf\{n \geq 1 : \mathcal{F}^n(z) \in M\}. \tag{4}$$

Let $M_n = \{R = n\} \subset M$ be the n-th level set of R, for each $n \geq 1$. Moreover, for $n, m \geq 1$, we denote

$$p_{n,m} := \frac{\nu(F^{-1} M_m \cap M_n)}{\nu(M_n)} \tag{5}$$

The quantities $p_{n,m}$ can be thought of as the transition probability of going from cell M_n to cell M_m in one iteration. It is important to note that everything in M_n with $n \geq 3$ must go to M_1 if the neighbourhood is sufficiently small. From M_2, although it cannot go to cells of higher indices, it is possible, however, to go back to itself because of the presence of period-four-orbit-like trajectories. There is a positive probability to go from M_1 to any cells. Fig. 2 shows the structure of level sets of R in the phase space corresponding to Γ_1 and Γ_2, respectively, for the first table in Fig 1.

Now consider the induced collision map $F : M \to M$ given by: $F(z) = \mathcal{F}^{R(z)}(z)$. The function F is discontinuous on the lines separating the cells M_n's. Moreover, F preserves the conditional measure ν on M, where for each $B \subset M$, $\nu(B) := \frac{\mu(B)}{\mu(M)}$. The map $F : M \to M$ is strongly hyperbolic and has exponential decay of correlations.

Since the set M is partitioned by the cells M_n's, we also have a partition for $\mathcal{M}$:

$$\mathcal{M} = \cup_{n=1}^{\infty} \cup_{k=0}^{n-1} \mathcal{F}^k M_n.$$

An element $z \in \mathcal{M}$ can be represented by the pair (y, i) where $\Pi(z) = y$ is the projection onto the base M and $z = \mathcal{F}^i(y)$ with $0 \leq i \leq R(y) - 1$. Let $\mathcal{F}_0^{\mathcal{M}}$ be the σ-algebra generated by this partition of $\mathcal{M}$. We now state the main theorem of this paper:

Theorem 1 *Let $\mathcal{D}$ be the billiard table with flat points. Let $f : \mathcal{M} \to \mathbf{R}$ be a bounded $\mathcal{F}_0^{\mathcal{M}}$-measurable function and $\mu(f) = 0$. Then we have:*

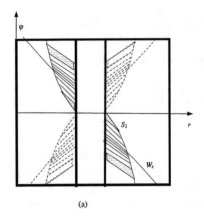
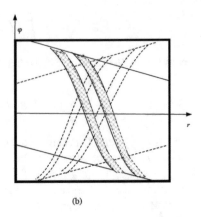

(a)   (b)

Fig. 2 **a** The level sets of R in the phase space corresponding to collisions on Γ_1. The shadowed region is M_2. The dashed curves are the boundary of level curves for $R \circ F$. W_s is the weak stable manifold for the periodic-2 trajectory. **b** The level sets for R and $R \circ F$ in the phase space corresponding to collisions on Γ_2. Note that in this phase space, besides M_1 the level sets of R only contains two components of M_2. The dashed lines are boundaries of FM_2

$$\lim_{n\to\infty} \mu \left\{ \frac{S_n f}{\sqrt{n}} \leq t \right\} = \frac{1}{\sqrt{2\pi}\sigma_f} \int_{-\infty}^{t} e^{-\frac{s^2}{2\sigma_f^2}} \, ds. \tag{6}$$

for all $-\infty < t < \infty$. *Here:*

$$S_n f = f + f \circ \mathcal{F} + \cdots + f \circ \mathcal{F}^{n-1}.$$

Moreover, $\sigma_f^2 = \sigma_{\tilde{f}}^2 \mu(M)$, *where* $\sigma_{\tilde{f}}^2$ *is given in Theorem 2.*

Remark 1 Since $R \in L^{2+\delta}$ with $\delta > 0$ (see Lemma 1 below), the bounded condition on f can actually be replaced by $f \in L^{2+2/\delta}(\mathcal{M}, \mu)$, see [8].

2 Induced Function

In order to prove Theorem 1, we will first prove that the induced function of f also satisfies a central limit theorem. The induced function of f is given by:

$$\tilde{f} := f + f \circ \mathcal{F} + \cdots + f \circ \mathcal{F}^{R-1}.$$

Lemma 1 *We have that* $R \in L^{2+\delta}(M, \nu)$ *for any* $0 < \delta < a - 1$.

Proof The verification of this lemma is straightforward, since:

$$v(R > n) \leq C' \cdot n^{-a-1} \tag{7}$$

for every $n \geq 1$ and some uniform constant C' (see [4]). We recall that $a = \frac{\beta+2}{\beta-2} > 1$. $\qquad\square$

Suppose that $f(z) : \mathcal{M} \to \mathbf{R}$ is $\mathcal{F}_0^{\mathcal{M}}$-measurable. Then one can check that $\tilde{f}$ is constant on each cell M_n and furthermore $\tilde{f} \in L^2(M, \nu)$ since $f \in L^\infty(\mathcal{M}, \mu)$.

Theorem 2 (CLT for the induced function) *Let $f : \mathcal{M} \to \mathbf{R}$ be defined as in Theorem 1 and $\tilde{f}$ its induced function on M. Then we have*

$$\lim_{n \to \infty} \nu \left\{ \frac{S_n \tilde{f} - n\nu(\tilde{f})}{\sqrt{n}} \leq t \right\} = \frac{1}{\sqrt{2\pi}\sigma_{\tilde{f}}} \int_{-\infty}^t e^{-\frac{s^2}{2\sigma_{\tilde{f}}^2}} ds. \tag{8}$$

for all $-\infty < t < \infty$, where $S_n \tilde{f} = \tilde{f} + \tilde{f} \circ F + \cdots + \tilde{f} \circ F^{n-1}$. and

$$\sigma_{\tilde{f}}^2 = Var(\tilde{f}) - 2(\mathbf{E}(\tilde{f}|M_1))^2 \nu(M_1) + 2\frac{p_{2,2}}{p_{1,2}} \mathbf{E}(\tilde{f}|M_2)\nu(M_2)\big(\mathbf{E}(\tilde{f}|M_1)(p_{1,1} - 1) +$$

$$+ \mathbf{E}(\tilde{f}|M_2)p_{1,2} - \mathbf{E}(\tilde{f} \circ F|M_1)\big).$$

An important special case of Theorem 2 is when $\tilde{f}$ is the return time function:

Corollary 1 *Let f be defined by:*

$$f(z) = \begin{cases} 1 & \text{if } z \in \mathcal{M} \setminus M \\ 1 - \nu(R) & \text{if } z \in M, \end{cases} \tag{9}$$

then $\tilde{f} = R - \nu(R)$. Thus the (centralised) return time function $R - \nu(R)$ also satisfies the central limit theorem, that is:

$$\lim_{n \to \infty} \nu \left\{ \frac{S_n R - n\nu(R)}{\sqrt{n}} \leq t \right\} = \frac{1}{\sqrt{2\pi}\sigma_R} \int_{-\infty}^t e^{-\frac{s^2}{2\sigma_R^2}} ds. \tag{10}$$

for all $-\infty < t < \infty$, with

$$\sigma_R^2 = Var(R) - 2(1 - \nu(R))^2 \nu(M_1) + \frac{p_{2,2}}{p_{1,2}}(4 - 2\nu(R))\nu(M_2) \times$$

$$\times \left(p_{1,2} - p_{1,3} + \nu(R)(1 + p_{1,3}) - \mathbf{E}(R \circ F|M_1) \right).$$

Assuming Theorem 2, we now show that the Theorem 1 is true. This standard result is proved in several references, for example, [1, 3]. For completeness, we give a proof here. But before we go to the proof of this lemma, we need some basic results.

Lemma 2 *For each $n \geq 1$, let $n_x(n)$ be the number of times the point mass comes back to M during the first n iterations. Then for v-a.e. $x \in M$ we have:*

$$\lim_{n \to \infty} \frac{n}{n_x(n)} = v(R).$$

Proof We first note that, for v-almost every x, $n_x(n) \to \infty$ as $n \to \infty$. The set of x such that the sequence $\{n_x(n)\}$ is bounded has measure 0: it is the countable union of all preimages of the set $\{R = \infty\}$.

The induced map F is ergodic and $R \in L^1(M, v)$, therefore we have, by Birkhoff ergodic theorem:

$$\lim_{n \to \infty} \frac{S_n R}{n} = v(R)$$

for almost every $x \in M$. For such an $x \in M$, since $S_{n_x(n)} R \leq n < S_{n_x(n)+1} R$, we have that:

$$\frac{S_{n_x(n)} R}{n_x(n)} \leq \frac{n}{n_x(n)} \leq \frac{S_{n_x(n)+1} R}{n_x(n) + 1} \cdot \frac{n_x(n) + 1}{n_x(n)}.$$

Therefore we have for almost every $x \in M$ that

$$\lim_{n \to \infty} \frac{n}{n_x(n)} = v(R). \qquad \square$$

Corollary 2 *We have:*

$$\lim_{n \to \infty} v \left(\frac{n_x(n) - n/v(R)}{\sqrt{n}} \leq t \right) = \frac{1}{\sqrt{2\pi}\sigma} \int_{-\infty}^{t} e^{-\frac{s^2}{2\sigma^2}} ds.$$

for all $t \in (-\infty, \infty)$ and $\sigma^2 = \sigma_R^2/(v(R))^3 = \sigma_R^2 \mu(M)^3$.

Note that we used the Kac formula $v(R)^{-1} = \mu(M)$ in the above Corollary.

Lemma 3 *Theorem 2 implies Theorem 1.*

Proof In this proof, we will assume for simplicity that the function f is bounded. See [6], Appendix A for a similar but longer proof of the more general case. Without loss of generality, assume that $\mu(f) = 0$ and therefore we also have $v(\tilde{f}) = 0$. Let $m = m(n) = \lfloor n/v(R) \rfloor$. Corollary 2 implies that for any $\varepsilon > 0$, there exists $A_\varepsilon > 0$ such that

$$v(|n_x - m| \geq A\sqrt{n}) \leq \varepsilon.$$

First we prove that with respect to μ on $\mathcal{M}$ we have:

$$\frac{S_n f \circ \Pi}{\sqrt{n}} \implies N(0, \sigma_f^2).$$

We have:

$$\frac{\mathcal{S}_n f}{\sqrt{n}} = \frac{\mathcal{S}_{n_1}\tilde{f}}{\sqrt{n}} + \frac{\mathcal{S}_{n_2}\tilde{f} - \mathcal{S}_{n_1}\tilde{f}}{\sqrt{n}} + \frac{\mathcal{S}_n f - \mathcal{S}_{n_2}\tilde{f}}{\sqrt{n}}.$$

The first term converges to $N(0, \sigma_{\tilde{f}}^2)$ with respect to ν by our assumption and $\sigma_f^2 = \sigma_{\tilde{f}}^2/\nu(R)$. The second and third terms converge to 0 in probability, by Birkhoff ergodic theorem and the fact that f is a bounded function. Thus we have shown that on (M, ν):

$$\frac{\mathcal{S}_n f}{\sqrt{n}} \implies N(0, \sigma_f^2). \tag{11}$$

We define a new probability measure ξ on M by $d\xi = R/\nu(R)d\nu$. Since $\xi \ll \nu$, the central limit theorem (11) also holds with respect to ξ. We have:

$$\int_{\mathfrak{M}} exp\left(it\frac{\mathcal{S}_n f \circ \Pi}{\sqrt{n}}\right) d\mu = \int_M R \exp\left(it\frac{\mathcal{S}_n f}{\sqrt{n}}\right) d\mu = \int_M \frac{R}{\nu(R)} \exp\left(it\frac{\mathcal{S}_n f}{\sqrt{n}}\right) d\nu. \tag{12}$$

This shows that on $(\mathfrak{M}, \mu)$:

$$\frac{\mathcal{S}_n f \circ \Pi}{\sqrt{n}} \implies N(0, \sigma_f^2).$$

To complete the prove of this lemma, we will show that $\frac{\mathcal{S}_n f}{\sqrt{n}} - \frac{\mathcal{S}_n f \circ \Pi}{\sqrt{n}} \longrightarrow 0$ in probability.

$$\mathcal{S}_n f(y, i) - \mathcal{S}_n f(y, 0) = \sum_{k=0}^{n-1} f \circ \mathcal{F}^k(y, i) - \sum_{k=0}^{n-1} f \circ \mathcal{F}^k(y, 0)$$

$$= -\sum_{k=0}^{i-1} f \circ \mathcal{F}^k(y, 0) + \sum_{k=n}^{n+i-1} f \circ \mathcal{F}^k(y, 0)$$

$$= -\sum_{k=0}^{i-1} f(y, k) + \sum_{k=0}^{i-1} f \circ \mathcal{F}^n(y, k).$$

Since $|\mathcal{S}_n f(y, i) - \mathcal{S}_n f(y, 0)| \leq 2\|f\|_\infty R$, we have that

$$\frac{\mathcal{S}_n f}{\sqrt{n}} - \frac{\mathcal{S}_n f \circ \Pi}{\sqrt{n}} \longrightarrow 0 \text{ in probability.}$$

Thus we have shown that Theorem 2 implies Theorem 1. $\qquad\qquad\square$

3 Central Limit Theorem for the Induced Function

We devote this section to prove a central limit theorem on the induced system (M, F, ν) of which Theorem 2 is a special case:

Theorem 3 *Let $X : M \to \mathbf{R}$ be an $\mathcal{F}_0$-measurable function such that $X \in L^2(M, \nu)$ and $\mathbf{E}(X) = 0$. Then*

$$\frac{S_n X}{\sqrt{n}} \Rightarrow N(0, \sigma_X^2), \tag{13}$$

where the variance σ_X^2 is given by formula (31).

There is a filtration of σ-algebras on M:

$$\mathcal{F}_n = \sigma(R \circ F^k : -n \leq k \leq n) \tag{14}$$

for $n \geq 0$ and $\mathcal{F}_n = \{\emptyset, M\}$ for $n < 0$. Let $X_n = X \circ F^n$ for $n \geq 0$. Because F preserves the probability measure ν, the sequence $\{X_n\}_{n \geq 0}$ is a stationary stochastic process adapted to the filtration $\{\mathcal{F}_n\}$. By replacing X by $X - \mathbf{E}(X)$, we can assume that $\mathbf{E}(X) = 0$.

Our method in proving that $\frac{S_n X}{\sqrt{n}}$ converges to a normal distribution as $n \to \infty$ is to approximate the Birkhoff sum by a series of martingale differences for which a central limit theorem is already proved, see [7]:

Lemma 4 *Let $\{Z_j : j \geq 1\}$ be a stationary ergodic sequence of martingale differences such that $\mathbf{E}(Z_1^2) = \sigma^2 < \infty$. Then we have*

$$\frac{S_n Z}{\sqrt{n}} \Longrightarrow N(0, \sigma^2).$$

The convergence here is in distribution.

Our approximation is as follows. Fix any large integer $k \geq 1$. Then for any $n \geq 1$ we have a decomposition:

$$X_n = \mathbf{E}(X_n | \mathcal{F}_{n-k}) + h_k \circ F^{n-1} + u_n^k - v_n^k, \tag{15}$$

where $h_k = \sum_{i=1}^{k} \left(\mathbf{E}(X_i | \mathcal{F}_1) - \mathbf{E}(X_i | \mathcal{F}_0) \right)$, $v_{n-1}^k = u_n^k$ and

$$u_n^k = \sum_{i=0}^{k-2} \left(\mathbf{E}(X_{n+i} | \mathcal{F}_{n-1}) - \mathbf{E}(X_{n+i} | \mathcal{F}_{n-k+i}) \right). \tag{16}$$

Therefore:

$$X_0 + \cdots + X_{n-1} = \sum_{i=0}^{n-2} h_k \circ F^i + \mathbf{E}(X_0 + \cdots + X_{k-1}|\mathcal{F}_0) - v_{n-1}^k + \sum_{i=k}^{n-1} \mathbf{E}(X_i|\mathcal{F}_{i-k}).$$

(17)

Note that $\mathbf{E}(X_i|\mathcal{F}_{i-k}) = \mathbf{E}(X_k|\mathcal{F}_0) \circ F^{i-k}$ for $i \geq k$, and $v_n^k = v_k^k \circ F^{n-k}$. We have:

$$\limsup_{n\to\infty} n^{-1} \mathbf{E}\big(X_0 + \cdots + X_{n-1} - \sum_{i=0}^{n-2} h_k \circ F^i\big)^2$$

$$= \limsup_{n\to\infty} n^{-1} \mathbf{E}\Big(\mathbf{E}(X_0 + \cdots + X_{k-1}|\mathcal{F}_0) - v_{n-1}^k + \sum_{i=k}^{n-1} \mathbf{E}(X_i|\mathcal{F}_{i-k})\Big)^2$$

$$\leq 3 \limsup_{n\to\infty} n^{-1} \mathbf{E}\Big(\sum_{i=k}^{n-1} \mathbf{E}(X_i|\mathcal{F}_{i-k})\Big)^2$$

$$= 3 \limsup_{n\to\infty} \Big(\frac{n-k}{n} \mathbf{E}\big(\mathbf{E}(X_k|\mathcal{F}_0)\big)^2 + \frac{2}{n} \sum_{i=1}^{n-k-1} (n-k-i)\mathbf{E}\big(\mathbf{E}(X_k|\mathcal{F}_0) \cdot \mathbf{E}(X_k|\mathcal{F}_0) \circ F^i\big)\Big).$$

Since $X : M \to \mathbf{R}$ is an $\mathcal{F}_0$-measurable function, we can compute the quantities $\mathbf{E}(X_k|\mathcal{F}_0)$ rather explicitly.

Lemma 5 Let $\mathbf{E}(X_k|\mathcal{F}_0) = \sum_{n=1}^{\infty} a_n^{(k)} \chi_{M_n}$, where $a_n^{(k)} = \mathbf{E}(X_k|M_n)$ for $n \geq 1$ and $k \geq 0$. We have a recurrence relation:

$$a_i^{(k+1)} = \sum_{m=1}^{\infty} a_m^{(k)} p_{i,m} \text{ for } i = 1, 2, \text{ and } a_n^{(k+1)} = a_1^{(k)} \text{ for } k \geq 0 \text{ and } n \geq 3. \quad (18)$$

Moreover,

$$\lim_{k\to\infty} a_i^{(k)} = 0, \text{ for } i = 1, 2. \quad (19)$$

Proof Suppose that $\mathbf{E}(X_k|\mathcal{F}_0) = \sum_{n=1}^{\infty} a_n^{(k)} \chi_{M_n}$. Then

$$\mathbf{E}(X_{k+1}|\mathcal{F}_1) = \mathbf{E}(X_k|\mathcal{F}_0) \circ F = \sum_{n=1}^{\infty} a_n^{(k)} \chi_{F^{-1}M_n},$$

and thus:

$$\mathbf{E}(X_{k+1}|\mathcal{F}_0) = \sum_{n=1}^{\infty} \mathbf{E}(\mathbf{E}(X_{k+1}|\mathcal{F}_1)|M_n) \chi_{M_n} = \sum_{n=1}^{\infty} \Big(\sum_{m=1}^{\infty} a_m^{(k)} p_{n,m}\Big) \chi_{M_n},$$

where we define $p_{n,m}$ as in (5). for $n, m \geq 1$. It is straightforward that $\sum_{m=1}^{\infty} p_{nm} = 1$ for any $n \geq 1$ since the cells M_n's are disjoint and the map F is invertible. Suppose that $x \in M_n$; that means the point mass will enter the neighbourhood of the flat points and come out after $n - 1$ collisions with the boundary. For $n \geq 3$, by shrinking the neighbourhood if necessary, once the point mass come out it will not come back to the neighbourhood after at least 2 collisions with the good part of the boundary of the table. That is to say $F^{-1}M_1 \cap M_n = M_n$, hence $p_{n,1} = 1$, for $n \geq 3$. In essence, we have a three-state Markov chain. Therefore we have that:

$$\mathbf{E}(X_{k+1}|\mathcal{F}_0) = \left(\sum_{m=1}^{\infty} a_m^{(k)} p_{1,m} \right) \chi_{M_1} + \left(\sum_{m=1}^{\infty} a_m^{(k)} p_{2,m} \right) \chi_{M_2} + a_1^{(k)} \sum_{n=3}^{\infty} \chi_{M_n}. \quad (20)$$

Let $z_k = (a_1^{(k)}, a_2^{(k)}, a_1^{(k-1)})^t$, and

$$(A_{ij}) = \begin{pmatrix} p_{1,1} & p_{1,2} & 1 - p_{1,1} - p_{1,2} \\ p_{2,1} & 1 - p_{2,1} & 0 \\ 1 & 0 & 0 \end{pmatrix}. \quad (21)$$

The recurrence can then be written in matrix form as:

$$z_{k+1} = A z_k \quad \text{for } k \geq 1; \quad z_1 = (a_1^{(1)}, a_2^{(1)}, a_1^{(0)})^t. \quad (22)$$

We note that the first row of A is strictly positive, thus A is an irreducible, aperiodic stochastic matrix and the unique stationary probability vector is $\pi = (\nu(M_1), \nu(M_2), \nu(M_{n \geq 3}))$:

$$\nu(M_1) A_{12} = \nu(M_2) A_{21}$$
$$\nu(M_1) A_{13} = \nu(M_{n \geq 3})$$

It follows that $\lim_{k \to \infty} a_i^{(k)} = \pi \cdot z_1$ for $i = 1, 2$. Furthermore, $\pi \cdot z_1 = \mathbf{E}(\mathbf{E}(X_1|\mathcal{F}_0)) = 0$. Thus we have:

$$\lim_{k \to \infty} a_i^{(k)} = 0, \text{ for } i = 1, 2. \qquad \square$$

Lemma 6

$$\lim_{k \to \infty} \mathbf{E}(\mathbf{E}(X_k|\mathcal{F}_0))^2 = 0.$$

Proof We recall that

$$\mathbf{E}(X_k|\mathcal{F}_0) = \sum_{n \geq 1} a_n^{(k)} \chi_{M_n} = a_1^{(k)} \chi_{M_1} + a_2^{(k)} \chi_{M_2} + a_1^{(k-1)}(1 - \chi_{M_1} - \chi_{M_2}).$$

Therefore:

$$\mathbf{E}\left(\mathbf{E}(X_k|\mathcal{F}_0)\right)^2 = (a_1^{(k)})^2 \nu(M_1) + (a_2^{(k)})^2 \nu(M_2) + (a_1^{(k-1)})^2 (1 - \nu(M_1) - \nu(M_2)).$$

Thus we have:

$$\lim_{k\to\infty} \mathbf{E}\,(\mathbf{E}(X_k|\mathcal{F}_0))^2 = 0. \qquad \qquad \square$$

Lemma 7

$$\lim_{k\to\infty} \sum_{i=1}^{\infty} \mathbf{E}\big(\mathbf{E}(X_k|\mathcal{F}_0) \cdot \mathbf{E}(X_k|\mathcal{F}_0) \circ F^i\big) = 0.$$

Proof As before we have

$$\mathbf{E}(X_k|\mathcal{F}_0) \cdot \mathbf{E}(X_{k+i}|\mathcal{F}_0) = a_1^{(k)} a_1^{(k+i)} \chi_{M_1} + a_2^{(k)} a_2^{(k+i)} \chi_{M_2} + a_1^{(k-1)} a_1^{(k+i-1)}(1 - \chi_{M_1} - \chi_{M_2}).$$

Taking the expectation we have:

$$\begin{aligned}
\mathbf{E}(\mathbf{E}(X_k|\mathcal{F}_0) \cdot \mathbf{E}(X_{k+i}|\mathcal{F}_0)) &= a_1^{(k)} a_1^{(k+i)} \nu(M_1) + a_2^{(k)} a_2^{(k+i)} \nu(M_2) + \\
&\quad + a_1^{(k-1)} a_1^{(k+i-1)}(1 - \nu(M_1) - \nu(M_2)) \\
&= a_1^{(k)} \nu(M_1)(a_1^{(k+i)} - a_1^{(k+i-1)}) + a_2^{(k)} \nu(M_2)(a_2^{(k+i)} - a_1^{(k+i-1)}).
\end{aligned}$$

To deal with the last term, we have for $n \geq 2$ that:

$$a_2^{(n)} - a_1^{(n-1)} = (a_2^{(n-1)} - a_1^{(n-1)})A_{22} \qquad \qquad (23)$$

$$a_2^{(n-1)} - a_1^{(n-1)} = \frac{a_1^{(n)} - a_1^{(n-1)}}{A_{12}} + \frac{(a_1^{(n-1)} - a_1^{(n-2)})A_{13}}{A_{12}}. \qquad (24)$$

Thus the series $\sum_{i=1}^{\infty} \mathbf{E}(\mathbf{E}(X_k|\mathcal{F}_0) \cdot \mathbf{E}(X_k|\mathcal{F}_0) \circ F^i)$ is in fact a telescoping series and noting that $a_i^{(k)} \to 0$ as $k \to \infty$ for $i = 1, 2$, it must be the case that:

$$\lim_{k\to\infty} \sum_{i=1}^{\infty} \mathbf{E}(\mathbf{E}(X_k|\mathcal{F}_0) \cdot \mathbf{E}(X_k|\mathcal{F}_0) \circ F^i) = 0. \qquad \square$$

Thus for any positive sequence $\varepsilon_k \to 0$ as $k \to \infty$, there exists a sequence $n_k \to \infty$ as $k \to \infty$ such that

$$\limsup_{n\to\infty} n^{-1}\mathbf{E}\big(X_0 + \cdots + X_{n-1} - \sum_{i=0}^{n-2} h_{n_k} \circ F^i\big)^2 < \varepsilon_k.$$

The sequence $\{h_{n_k} \circ F^i\}_{i\geq 0}$ is a stationary sequence of martingale differences adapted to the filtration $\{\mathcal{F}_i\}$. The CLT holds for this sequence:

$$n^{-1/2} \sum_{i=0}^{n-1} h_{n_k} \circ F^i \implies N(0, \sigma_k^2) \qquad \qquad (25)$$

where $\sigma_k^2 = \mathbf{E}(h_{n_k}^2)$.

Next, we show that the sequence $\{\sigma_k\}$ converges to some limit as $k \to \infty$.

$$
\begin{aligned}
(\sigma_i - \sigma_j)^2 &\leq \mathbf{E}\left(h_{n_i} - h_{n_j}\right)^2 \\
&= n^{-1}\mathbf{E}\left(\sum_{m=0}^{n-1}(h_{n_i} - h_{n_j}) \circ F^m\right)^2 \\
&\leq 2(\varepsilon_i + \varepsilon_j).
\end{aligned}
$$

Therefore $\{\sigma_k\}$ is a Cauchy sequence and hence $\sigma_k \to \sigma_X$ as $k \to \infty$ for some constant σ_X and

$$
\lim_{n \to \infty} \mathbf{E}\left(\frac{S_n X}{\sqrt{n}}\right)^2 = \sigma_X^2.
$$

Finally, the variance σ_X^2 can be computed directly as below:

For any $n \geq 1$:

$$
\begin{aligned}
Cov(X, X \circ F^n) &= \mathbf{E}\left(\mathbf{E}(X \circ F^n | \mathcal{F}_0) \cdot X\right) \\
&= \mathbf{E}\left(\left(a_1^{(n)}\chi_{M_1} + a_2^{(n)}\chi_{M_2} + a_1^{(n-1)}\sum_{m \geq 3}\chi_{M_m}\right) \cdot \sum_{m=1}^{\infty}a_m^{(0)}\chi_{M_m}\right) \\
&= a_1^{(n)}a_1^{(0)}v(M_1) + a_2^{(n)}a_2^{(0)}v(M_2) + a_1^{(n-1)}\sum_{m=3}^{\infty}a_m^{(0)}v(M_m) \\
&= (a_1^{(n)} - a_1^{(n-1)})a_1^{(0)}v(M_1) + (a_2^{(n)} - a_1^{(n-1)})a_2^{(0)}v(M_2).
\end{aligned}
$$

In particular, for $n = 1$:

$$
\begin{aligned}
Cov(X, X \circ F) &= (a_1^{(1)} - a_1^{(0)})a_1^{(0)}v(M_1) + (a_2^{(1)} - a_1^{(0)})a_2^{(0)}v(M_2) \qquad (26) \\
&= (a_1^{(1)} - a_1^{(0)})a_1^{(0)}v(M_1) + (a_2^{(0)} - a_1^{(0)})A_{22}a_2^{(0)}v(M_2). \qquad (27)
\end{aligned}
$$

For $n \geq 2$, we have:

$$
a_2^{(n)} - a_1^{(n-1)} = (a_2^{(n-1)} - a_1^{(n-1)})A_{22} \qquad (28)
$$

$$
a_2^{(n-1)} - a_1^{(n-1)} = \frac{a_1^{(n)} - a_1^{(n-1)}}{A_{12}} + \frac{(a_1^{(n-1)} - a_1^{(n-2)})A_{13}}{A_{12}}. \qquad (29)
$$

Therefore:

$$Cov(X, X \circ F^n) = (a_1^{(n)} - a_1^{(n-1)})a_1^{(0)}v(M_1)$$

$$+ \left(a_1^{(n)} - a_1^{(n-1)} + (a_1^{(n-1)} - a_1^{(n-2)})A_{13}\right)\frac{A_{22}}{A_{12}}a_2^{(0)}v(M_2).$$

$$= (a_1^{(n)} - a_1^{(n-1)})(a_1^{(0)}v(M_1) + W) + (a_1^{(n-1)} - a_1^{(n-2)})A_{13}W,$$

where:

$$W = \frac{A_{22}}{A_{12}}a_2^{(0)}v(M_2). \qquad (30)$$

We can then compute the variance of $\frac{S_n X}{\sqrt{n}}$ as follows:

$$Var\left(\frac{S_n X}{\sqrt{n}}\right) = Var(X) + \frac{2}{n}\sum_{k=1}^{n-1}(n-k)Cov(X, X \circ F^k)$$

$$= Var(X) + 2\sum_{k=1}^{n-1}Cov(X, X \circ F^k) - \frac{2}{n}\sum_{k=1}^{n-1}kCov(X, X \circ F^k).$$

The second term is:

$$\sum_{k=1}^{n-1}Cov(X, X \circ F^k) = Cov(X, X \circ F) + (a_1^{(n-1)} - a_1^{(1)})(a_1^{(0)}v(M_1) + W)$$

$$+ (a_1^{(n-2)} - a_1^{(0)})A_{13}W.$$

Taking limit as $n \to \infty$, the third term converges to 0 by Kronecker's lemma or by direct verification. Thus we have:

$$\sigma_X^2 = \lim_{n\to\infty} Var\left(\frac{S_n X}{\sqrt{n}}\right) = Var(X) - 2(a_1^{(0)})^2 v(M_1) + 2W\left(a_1^{(0)}A_{11} + a_2^{(0)}A_{12} - a_1^{(0)} - a_1^{(1)}\right).$$

$$(31)$$

Thus we have shown that $\frac{S_n X}{\sqrt{n}} \implies N(0, \sigma_X^2)$ in distribution and completed the proof of Theorem 3.

Remark 2 Our method also works for functions X that are $\mathcal{F}_m$-measurable for any $m \geq 0$. The martingale approximation is virtually the same, and the estimations of the errors are easily reduced to estimation of the case X is $\mathcal{F}_0$-measurable since we are dealing with stationary stochastic sequences. Thus the central limit theorem actually holds for a much larger class of observables than those considered in Theorem 1. However, a drawback is that a formula for the diffusion constant would be more complicated.

References

1. Bálint, P., Chernov, N., Dolgopyat, D.: Limit theorems for dispersing billiards with cusps. Comm. Math. Phys. **308**, 479–510 (2011)
2. Bunimovich, L., Sinai, Y., Chernov, N.: Statistical properties of two-dimensional hyperbolic billiards. Russ. Math. Surv. **46**, 47–106 (1990)
3. Chernov, N., Markarian, R.: Chaotic Billiards. Mathematical Surveys and Monographs, vol. 127. AMS, Providence (2007)
4. Chernov, N., Zhang, H.-K.: A family of chaotic billiards with variable mixing rates. Stoch. Dyn. **5**, 535–553 (2005)
5. Chernov, N., Zhang, H.-K.: Billiards with polynomial mixing rates. Nonlinearity **18**, 1527–1553 (2005)
6. Gouëzel, S.: Statistical properties of a skew product with a curve of neutral points. Ergod. Theory Dyn. Syst. **27**, 123–151 (2007)
7. Hall, P., Heyde, C.C.: Martingale Limit Theorem and Its Application. Academic Press, New York (1980)
8. Melbourne, I., Török, A.: Statistical limit theorems for suspension flows. Isr. J. Math. **144**, 191–209 (2004)
9. Mohr, L., Zhang, H-K.: Superdiffusions for Certain Nonuniformly Hyperbolic Systems (submitted, 2017)

Almost Sure Rates of Mixing for Random Intermittent Maps

Marks Ruziboev

Abstract We consider a family $\mathcal{F}$ of maps with two branches and a common neutral fixed point 0 such that the order of tangency at 0 belongs to some interval $[\alpha_0, \alpha_1] \subset (0, 1)$. Maps in $\mathcal{F}$ do not necessarily share common Markov partition. At each step a member of $\mathcal{F}$ is chosen independently with respect to the uniform distribution on $[\alpha_0, \alpha_1]$. We show that the construction of the random tower in Bahsoun et al. (Quenched Decay of Correlations for Slowly Mixing Systems, 2018, [5]) with *general return time* can be carried out for random compositions of such maps. Thus their general results are applicable and gives upper bounds for the quenched decay of correlations of form the $n^{1-1/\alpha_0+\delta}$ for the any $\delta > 0$.

Keywords Random dynamical system · Quenched decay of correlations
Random induced schemes · General return times · Intermittent maps

1 Introduction

In recent years there has been remarkable interest in studying statistical properties of random dynamical systems induced by random compositions of different maps (see for example [1–6, 10, 11, 14, 15, 17] and references therein). In [4] i.i.d. random compositions of two Liverani–Saussol–Vaienti (LSV)[1] maps were considered and it was shown that the rate of decay of the annealed (averaged over all realisations) correlations is given by the fast dynamics. Recently the general results on quenched

---

[1] A subclass of the so called Pomeau–Manneville maps introduced in [18], and popularised by Liverani, Saussol and Vaienti in [16]. Such systems have attracted the attention of both mathematicians and physicists (see [15] for a recent work in this area).

Dedicated to Abdulla Azamov and Leonid Bunimovich on the occasion of their 70th birthday.

M. Ruziboev (✉)
Department of Mathematical Sciences, Loughborough University, Loughborough,
Leicestershire LE11 3TU, UK
e-mail: M.Ruziboev@lboro.ac.uk

© Springer Nature Switzerland AG 2018
A. Azamov et al. (eds.), *Differential Equations and Dynamical Systems*,
Springer Proceedings in Mathematics & Statistics 268,
https://doi.org/10.1007/978-3-030-01476-6_11

decay rates (i.e. decay rates for almost every realisation) for the random compositions of non-uniformly expanding maps were obtained in [5]. As an illustration it was shown ibidem that the general results are applicable to the random map induced by compositions of LSV maps with parameters in $[\alpha_0, \alpha_1] \subset (0, 1)$ chosen with respect to a suitable distribution ν on $[\alpha_0, \alpha_1]$. In the current note we fix the uniform distribution on $[\alpha_0, \alpha_1]$ and consider a family of maps with common neutral fixed point. Our maps do *not* share a common Markov partition. We show that the construction of the random tower of [5] with *general return time* can be carried out for the random compositions of such maps. Hence the main result of [5] is applicable. We obtain upper bounds for the quenched decay of correlations of the form $n^{1-1/\alpha_0+\delta}$ for any $\delta > 0$.

The paper is organised as follows. In Sect. 2, we give a formal definition of the family $\mathcal{F}$ and state the main result of the paper (Theorem 1). In Sect. 3, we construct uniformly expanding induced random map and show that the assumptions required in [5] are satisfied, i.e. we check uniform expansion, bounded distortion, decay rates for the tail of the return time and aperiodicity. Also we formulate a technical proposition in this section which is used to obtain the tail estimates and proved in Sect. 4.

2   The Set up and the Main Results

In this section we define the main object of the current note: the random maps. Fix two real numbers $0 < \alpha_0 < \alpha_1 < 1$. Let $I = [0, 1]$ and let $\mathcal{F}$ be a parametrised family of maps $T_\alpha : I \to I, \alpha \in [\alpha_0, \alpha_1]$ with the following properties.

(A1) There exists a C^1 function $x : [\alpha_0, \alpha_1] \to (0, 1)$, $\alpha \mapsto x_\alpha$ such that $T_\alpha :$ $[0, x_\alpha) \to [0, 1)$ and $T_\alpha : [x_\alpha, 1] \to [0, 1]$ are increasing diffeomorphisms.
(A2) $T'_\alpha(x) > 1$ for any $x > 0$.
(A3) There exists $\varepsilon_0 > 0$ and continuous functions $\alpha \mapsto c_\alpha$, $(x, \alpha) \mapsto f_\alpha(x)$ such that $f_\alpha(0) = 0$ and $T_\alpha(x) = x + c_\alpha x^{1+\alpha}(1 + f_\alpha(x))$ for any $x \in [0, \varepsilon_0]$.
(A4) Every T_α is C^3 on $(0, x_\alpha]$ with negative Schwarzian derivative.
(A5) $(x, \alpha) \mapsto T''_\alpha(x)$ and $(x, \alpha) \mapsto T'_\alpha(x)$ are continuous on $I \times [\alpha_0, \alpha_1]$.

Notice that the elements of $\mathcal{F}$ are parametrised according to the tangency near 0. Now, we describe the randomising dynamics. Let η be the normalised Lebesgue measure on $[\alpha_0, \alpha_1]$. Let $\Omega = [\alpha_0, \alpha_1]^{\mathbb{Z}}$ and $\mathbb{P} = \eta^{\mathbb{Z}}$. Then the shift map $\sigma : \Omega \to \Omega$ preservers $\mathbb{P}$, i.e. $\sigma_* \mathbb{P} = \mathbb{P}$. For $\omega \in \Omega$, $\omega = \ldots \omega_{-1}, \omega_0, \omega_1, \ldots$ let $\alpha(\omega) = \omega_0 \in [\alpha_0, \alpha_1]$. The *random map* is formed by random compositions of maps $T_{\alpha(\omega)} : I \to I$ from $\mathcal{F}$, where the compositions are defined as $T^n_\omega(x) = T_{\alpha(\sigma^{n-1}(\omega))} \circ \cdots \circ T_{\alpha(\omega)}(x)$. Below we use more shorter notation $T^n_\omega = T_{\omega_{n-1}} \circ \cdots \circ T_{\omega_0}(x)$. We are interested in studying the statistical properties of equivariant families of measures i.e. families of measures $\{\mu_\omega\}_{\omega \in \Omega}$ such that $(T_\omega)_* \mu_\omega = \mu_{\sigma \omega}$. Let μ be a probability measure on $I \times \Omega$ such that $\mu(A) = \int_\Omega \mu_\omega(A) d\mathbb{P}(\omega)$ for $A \subset I \times \Omega$. We say that the system $\{f_\omega, \mu_\omega\}_{\omega \in \Omega}$ (or simply $\{\mu_\omega\}_\omega$) is *mixing* if for all $\varphi, \psi \in L^2(\mu)$,

$$\lim_{n\to\infty}\left|\int_{\Omega}\int_0^1 \varphi_{\sigma^n\omega}\circ f_\omega^n\cdot\psi_\omega d\mu_\omega d\mathbb{P}-\int_{\Omega}\int_0^1\varphi_\omega d\mu_\omega d\mathbb{P}\int_{\Omega}\int_0^1\psi_\omega d\mu_\omega d\mathbb{P}\right|=0.$$

Further, future and past correlations are defined as follows. Let φ, $\psi : I \to \mathbb{R}$ be two observables on I. Then we define *future correlations* as

$$Cor_\mu^f(\varphi,\psi):=\left|\int(\varphi\circ T_\omega^n)\psi d\mu_{\sigma^n\omega}-\int\varphi d\mu_{\sigma^n\omega}\int\psi d\mu_\omega\right|$$

and *past correlations* as

$$Cor_\mu^p(\varphi,\psi):=\left|\int(\varphi\circ T_{\sigma^{-n}\omega}^n)\psi d\mu_\omega-\int\varphi d\mu_\omega\int\psi d\mu_{\sigma^{-n}\omega}\right|.$$

Theorem 1 *Let T_ω be the random map described above. Then for almost every $\omega\in\Omega$ there exists a family of absolutely continuous equivariant measures $\{\mu_\omega\}_\omega$ on I, which is mixing. Moreover, for every $\delta > 0$ there exists a full measure subset $\Omega_0\subset\Omega$ and a random variable $C_\omega : \Omega \to \mathbb{R}_+$ which is finite on Ω_0 such that for any $\varphi\in L^\infty(I)$, $\psi\in C^\eta(I)$ there exists a constant $C_{\varphi,\psi} > 0$ so that*

$$Cor_\mu^f(\varphi,\psi)\le C_\omega C_{\varphi,\psi}n^{1-\frac{1}{\alpha_0}+\delta}\text{ and }Cor_\mu^p(\varphi,\psi)\le C_\omega C_{\varphi,\psi}n^{1-\frac{1}{\alpha_0}+\delta}.$$

Furthermore, there exist constants $C > 0$, $u' > 0$ and $0 < v' < 1$ such that

$$P\{C_\omega > n\}\le Ce^{-u'n^{v'}}.$$

Remark 1 Notice that in the deterministic setting every mapping in the family $\mathcal{F}$ admits an absolutely continuous invariant probability measure, which is polynomially mixing at the rate $n^{1-1/\alpha}$ if $T_\alpha(x) = x + c_\alpha x^{1+\alpha}(1 + f_\alpha(x))$ (see [9, 20]). In the random setting the upper bounds we give are arbitrarily close to the sharp decay rates of the fastest mixing system in the family. Since the result holds for almost every $\omega\in\Omega$, and in principle there can be arbitrarily long compositions of systems in T_ω^n whose mixing rates are slower than that of T_{α_0} it is not expected that the mixing rate of the random system will be the same as the mixing rate of the fastest mixing system in the family $\mathcal{F}$ and C_ω integrable at the same time.

Remark 2 We also remark that we are choosing the family $\mathcal{F}$ so that all the maps in it share the common neutral fixed point 0. If we choose the family by allowing different maps having distinct neutral fixed points i.e. $T_\alpha(p(\alpha)) = p(\alpha)$, $T_\alpha'(p(\alpha)) = 1$ and $p(\alpha)\ne 0$ for a positive (with respect to η) measure set of parameters $\alpha\in[\alpha_0,\alpha_1]$ and expanding elsewhere, then the resulting random map is expanding on average. Whence one can apply spectral techniques as in [7] on the Banach space of quasi-Hölder functions from [13] to [19] and obtain exponential decay rates. Such systems are out of context in our setting since we are after systems with only polynomial decay of correlations.

To prove the theorem we construct a random induced map (or Random Young Tower) for T_ω with the properties described in [5]. Below we briefly recall the definition of induced map.

Let m denote the Lebesgue measure on I and $\Lambda \subset I$ be a measurable subset. We say T_ω admits a Random Young Tower with the base Λ if for almost every $\omega \in \Omega$ there exists a countable partition $\{\Lambda_j(\omega)\}_j$ of Λ and a return time function $R_\omega : \Lambda \to \mathbb{N}$ that is constant on each $\Lambda_j(\omega)$ such that

(P1) for each $\Lambda_j(\omega)$ the induced map $T_\omega^{R_\omega}|_{\Lambda_j(\omega)} : \Lambda_j(\omega) \to \Lambda$ is a diffeomorphism and there exists a constant $\beta > 1$ such that $(T_\omega^{R_\omega})' > \beta$.

(P2) There exists $\mathcal{D} > 0$ such that for all $\Lambda_j(\omega)$ and $x, y \in \Lambda_j(\omega)$

$$\left| \frac{(T_\omega^{R_\omega})' x}{(T_\omega^{R_\omega})' y} - 1 \right| \le \mathcal{D} \beta^{-s(T_\omega^{R_\omega}(x), T_\omega^{R_\omega}(y))},$$

where $s(x, y)$ is the smallest n such that $(T_\omega^{R_\omega})^n x$ and $(T_\omega^{R_\omega})^n y$ lie in distinct elements.

(P3) There exists $M > 0$ such that

$$\sum_n m\{x \in \Lambda \mid R_\omega(x) > n\} \le M \text{ for all } \omega \in \Omega;$$

There exist constants $C, u, v > 0, a > 1, b \ge 0$, a full measure subset $\Omega_1 \subset \Omega$, and a random variable $n_1 : \Omega_1 \to \mathbb{N}$ so that

$$\begin{cases} m\{x \in \Lambda \mid R_\omega(x) > n\} \le C \frac{(\log n)^b}{n^a}, & \text{whenever } n \ge n_1(\omega), \\ \mathbb{P}\{n_1(\omega) > n\} \le Ce^{-un^v}; \end{cases} \tag{1}$$

$$\int m\{x \in \Lambda \mid R_\omega = n\} d\mathbb{P}(\omega) \le C \frac{(\log n)^b}{n^{a+1}}. \tag{2}$$

(P4) There are $N \in \mathbb{N}$ and $\{t_i \in \mathbb{Z}_+ \mid i = 1, 2, ..., N\}$ such that g.c.d.$\{t_i\} = 1$ and $\varepsilon_i > 0$ so that for almost every $\omega \in \Omega$ and $i = 1, 2, \ldots N$ we have $m\{x \in \Lambda \mid R_\omega(x) = t_i\} > \varepsilon_i$.

Under the above assumptions it is proven in [5, Theorem 4.1] that there exists a family of absolutely continuous equivariant measures, which is mixing and the mixing rates have upper bound of form the $n^{1+\delta-a}$ for any $\delta > 0$ (Theorem 4.2, [5]). Therefore to prove Theorem 1 it is sufficient to construct an induced map $T_\omega^{R_\omega}$ with the properties (P1)–(P4), which is carried out in the next section.

3 Inducing Scheme

Here we will construct a uniformly expanding full branch induced random map on $\Lambda = (0, 1]$ for every $\omega \in \Omega$. Let $X_0(\omega) = 1$, $X_1(\omega) = x(\omega_0) = x_{\alpha(\omega)}$ and

$$X_n(\omega) = (T_\omega|_{[0, x(\omega_0))})^{-1} X_{n-1}(\sigma\omega) \text{ for } n \geq 2.$$

Let $I_n(\omega) = (X_n(\omega), X_{n-1}(\omega)]$. Then by definition $T_\omega(I_n(\omega)) = I_{n-1}(\sigma\omega)$. By induction we have

$$I_n(\omega) \xrightarrow{T_\omega} I_{n-1}(\sigma\omega) \xrightarrow{T_{\sigma\omega}} \cdots I_1(\sigma^{n-1}\omega) \xrightarrow{T_{\sigma^{n-1}\omega}} \Lambda.$$

Hence, every interval $I_n(\omega)$ first is mapped onto $I_1(\omega)$ and then is mapped onto Λ by the next iterate of T_ω. Define a return time $R_\omega : (0, 1] \to \mathbb{N}$ by setting $R_\omega|_{(X_n(\omega), X_{n-1}(\omega)]} = n$. Then the induced map $T_\omega^{R_\omega} : (0, 1] \to (0, 1]$ defined as $T_\omega^{R_\omega}|_{I_n(\omega)} = T_\omega^n$, for $n \geq 1$ is full branch. By assumptions (A1) and (A2) there exists $\beta > 1$ such that $(T_\omega^{R_\omega})' > \beta$ for all $\omega \in \Omega$. In fact, we can choose

$$\beta = \min_{\omega_0 \in [\alpha_0, \alpha_1]} \min_{x \in [x(\omega_0), 1]} |T'_{\omega_0}(x)|. \tag{3}$$

This proves (P1). By (A1) all the maps in $\mathcal{F}$ have two full branches with $x_\alpha < 1$. Hence, the interval where $R_\omega = 1$ has strictly positive length and thus (P4) is obviously satisfied.

To prove the remaining properties we use the following proposition, which is proved in Sect. 4.

Proposition 1 (1) *For every $\omega \in \Omega$ the sequence $\{X_n(\omega)\}_n$ is decreasing and $\lim_{n \to \infty} X_n(\omega) = 0$. Moreover, there exists a constant $C_0 > 0$ such that for all $\omega \in \Omega$*

$$\frac{1}{C_0 n^{1/\alpha_0}} \leq X_n(\omega) \leq \frac{C_0}{n^{1/\alpha_1}}. \tag{4}$$

(2) *There exist $C, u > 0$, $v \in (0, 1)$ and a random variable $n_1 : \Omega \to \mathbb{N}$ which is finite for $\mathbb{P}$-almost every $\omega \in \Omega$ such that*

$$\mathbb{P}\{\omega \mid n_1(\omega) > n\} \leq Ce^{-un^v}, \tag{5}$$

$$X_n(\omega) \leq Cn^{-1/\alpha_0}(\log n)^{1/\alpha_0} \quad \forall n \geq n_1, \tag{6}$$

$$\int (X_{n-1}(\omega) - X_n(\omega))d\mathbb{P}(\omega) \leq Cn^{-1-1/\alpha_0}(\log n)^{1/\alpha_0}. \tag{7}$$

Now we will prove (P3). For every $\omega \in \Omega$ by definition of R_ω and inequality (4) we have

$$m\{R_\omega > n\} = X_n(\omega) \leq C_0 n^{-1/\alpha_1}.$$

Since $\alpha_1 < 1$ we have $\sum n^{-1/\alpha_1} < +\infty$ and hence, there exists $M > 0$ such that

$$\sum_{n \geq 1} m\{R_\omega > n\} \leq M.$$

Inequalities (5) and (6) in Proposition 1 directly imply the inequalities in (1). Inequality (7) implies inequality (2) in (P3). It remains to show distortion estimates (P2) for the induced map. Our proof is based on Koebe principle. Recall that the Schwarzian derivative of a C^3 diffeomorphism g is defined as

$$Sg(x) = \frac{g'''(x)}{g'(x)} - \frac{3}{2}\left(\frac{g''(x)}{g'(x)}\right)^2.$$

It can be easily checked that if f and g are two maps such that $f' \geq 0$, $Sf < 0$ and $Sg \leq 0$, then $S(g \circ f) = (Sg) \circ f \cdot f' + Sf < 0$ i.e. the composition $g \circ f$ has negative Schwarzian derivative. We will use this observation in the proof of Lemma 1.

Let $J \subset J'$ be two intervals and let $\tau > 0$. J' is called a τ-scaled neighbourhood of J if both components of $J' \setminus J$ have length at least $\tau|J|$, where $|J|$ denotes the length of J. The Koebe principle [8, Chap. IV, Theorem 1.2] states that, if g is a diffeomorphism onto its image with $Sg < 0$, and $J \subset J'$ are two intervals such that $g(J')$ contains τ-scaled neighbourhood of $g(J)$ then there exists $\hat{K}(\tau)$ such that for any $x, y \in J$

$$\left|\frac{g'(x)}{g'(y)} - 1\right| \leq \hat{K}(\tau)\frac{|x-y|}{|J|}. \tag{8}$$

By applying the mean value theorem twice first in J and then in $(x, y) \subset J$ for any $x, y \in J$ we obtain

$$\frac{|g(x) - g(y)|}{|g(J)|} = \frac{|g'(v)|}{|g'(u)|}\frac{|x-y|}{|J|}$$

for some $u \in J$, $v \in (x, y)$. Now inequality (8) implies that $|g'(v)|/|g'(u)| \geq (1 + \hat{K}(\tau))^{-1}$. Thus

$$\left|\frac{g'(x)}{g'(y)} - 1\right| \leq K(\tau)\frac{|g(x) - g(y)|}{|g(J)|}, \tag{9}$$

for $K(\tau) = (1 + \hat{K}(\tau))\hat{K}(\tau)$.

Recall that by (A4) the left branch of T_ω has negative Schwarzian derivative for all $\omega \in \Omega$. This fact will be used in the proof of the following lemma.

Lemma 1 *There exists $K > 0$ such that for all $\omega \in \Omega$, $n \in \mathbb{N}$ and for $x, y \in I_n(\omega)$*

$$\left|\frac{(T_\omega^n)'(x)}{(T_\omega^n)'(y)} - 1\right| \leq K|T_\omega^n(x) - T_\omega^n(y)|.$$

Proof Notice, that $M = \max_{\omega_0 \in [\alpha_0, \alpha_1]} \max_{x \in I_1(\omega)} |T_\omega''(x)| < +\infty$ by (A5). Also, recall that $T_\alpha'|_{I_\omega} > \beta > 1$ for any $T_\alpha \in \mathcal{F}$. Thus for $n = 1$, we have

$$\left| \frac{(T_\omega)'(x)}{(T_\omega)'(y)} - 1 \right| \leq \frac{1}{\beta} |(T_\omega)'(x) - (T_\omega)'(y)| \leq \frac{M}{\beta^2} |T_\omega(x) - T_\omega(y)|.$$

For $n \geq 2$ we use Koebe principle mentioned above. Set $J = [X_n(\omega), X_{n-1}(\omega)]$ and $J' = [X_{n+1}(\omega), 2]$. We first extend $T_{\omega_{n-2}}, \ldots, T_{\omega_0}$ to $(0, +\infty)$ analytically, keeping the Schwarzian derivative non-positive.[2] Let $g = T_{\omega_{n-2}} \circ \cdots \circ T_{\omega_0}$. Then, g has negative Schwarzian derivative. We will show that $g(J')$ contains τ scaled neighbourhood of $g(J)$ for some $\tau > 0$, which is independent of ω. Since $g(X_n(\omega)) = X_1(\sigma^{n-1}\omega)$ and $g(X_{n+1}(\omega)) = X_2(\sigma^{n-1}\omega)$. It is sufficient to show that $X_1(\omega) - X_2(\omega)$ is bounded below by a constant independent of ω. By definition of X_n we have

$$|X_1(\omega) - X_2(\omega)| = |T_\omega^{-1}(1) - T_\omega^{-1} \circ T_{\sigma(\omega)}^{-1}(1)| \geq \frac{1}{\beta'} |1 - T_{\sigma(\omega)}^{-1}(1)| \geq \kappa > 0,$$

where $\beta' = \min\{T_\omega'(x) \mid (x, \omega_0) \in [\tilde{X}, 1] \times [\alpha_0, \alpha_1]\} > 1$ with $\tilde{X} = \min_\omega X_2(\omega)$ and $\kappa = \beta'(1 - \min_\alpha x_\alpha) > 0$ by (A1). Thus, using the fact $|g(J)| > 1 - \max_\alpha x_\alpha > 0$ from (9) we obtain

$$\left| \frac{g'(x)}{g'(y)} - 1 \right| \leq K |g(x) - g(y)|.$$

with $K = K(\tau)/(1 - \max_\alpha x_\alpha)$ which finished the proof.

Lemma 2 *There exists a constant $C > 0$ independent of ω such that for all $\omega \in \Omega$ and for any $x, y \in I_n(\omega)$*

$$\left| \log \frac{(T_\omega^{R_\omega})'(x)}{(T_\omega^{R_\omega})'(y)} \right| \leq C |T_\omega^{R_\omega}(x) - T_\omega^{R_\omega}(y)|.$$

Proof From now on we suppress the ω in R_ω, since no confusion arises. Note that $T_\omega^R(x)$ is the composition of the right branch of $T_{\sigma^{R-1}\omega}$ and g i.e. $T_\omega^R(x) = T_{\sigma^{R-1}\omega} \circ g(x)$. Therefore, by definition of β in (3) by Lemma 1 we have

$$\log \left| \frac{(T_\omega^R)'(x)}{(T_\omega^R)'(y)} \right| \leq K |T^R(x) - T^R(y)| + K |g(x) - g(y)| \leq K(1 + \frac{1}{\beta}) |T^R(x) - T^R(y)|.$$

Now, we will prove (P2). Together with an elementary inequality $|x - 1| \leq C |\log x|$ (for some $C > 0$, whenever $|\log x|$ is bounded above) Lemma 3 implies that for any $x, y \in I_n(\omega)$ we have

---

[2] Such extensions can be constructed easily. For example, for $f \in \mathcal{F}$ it is sufficient to take $\tilde{f}(x) = a(x - x_\alpha)^4 + b(x - x_\alpha)^3 + c(x - x_\alpha)^2 + d(x - x_\alpha) + 1$ with $a < bc/d$, where a, b, c are the Taylor coefficients of f at $x = x_\alpha$.

$$\left|\frac{(T_\omega^R)'x}{(T_\omega^R)'y} - 1\right| \le D(K, \beta)|T^R(x) - T^R(y)| \le \mathcal{D}\beta^{-s(T_\omega^R(x), T_\omega^R(y))},$$

where $\mathcal{D} = D(K, \beta)$ is a constant that depends only on K and β and the last inequality follows from the observation: if $x, y \in (0, 1]$ are such that $s(x, y) = n$ then $|x - y| \le \beta^{-n}$. Indeed, by definition $(T_\omega^R)^i(x)$ and $(T_\omega^R)^i(y)$ belong to the same element of the partition $\{I_k(\omega)\}$ for all $i = 0, ..., n - 1$. Thus by the mean value theorem

$$|x - y| = |[(T_\omega^R)^n]'(\xi)|^{-1}|(T_\omega^R)^n(x) - (T_\omega^R)^n(y)| \le \beta^{-n}.$$

4 The Proof of Proposition 1

We start by proving an auxiliary lemma, which is used in the proof.

Lemma 3 *For any $k \in \mathbb{N}$, $c \ge 1$ and $t > 0$ we have*

$$E_{\mathbb{P}}[e^{-(c\alpha(\sigma^k\omega)-\alpha_0)t}] = \frac{1}{\alpha_1 - \alpha_0}\frac{e^{\alpha_0 t(1-c)}}{ct}(1 - e^{-ct(\alpha_1-\alpha_0)}).$$

Proof Since σ preserves $\mathbb{P}$ we have

$$E_{\mathbb{P}}[e^{-(c\alpha(\sigma^k\omega)-\alpha_0)t}] = E_{\mathbb{P}}[e^{-(c\alpha(\omega)-\alpha_0)t}]$$
$$= \frac{1}{\alpha_1 - \alpha_0}\int_{\alpha_0}^{\alpha_1} e^{-(cx-\alpha_0)t}dx = \frac{1}{\alpha_1 - \alpha_0}\frac{e^{\alpha_0 t(1-c)}}{ct}(1 - e^{-ct(\alpha_1-\alpha_0)}).$$

Proof (Proof of Proposition 1) First we prove item (1). The first two assertions are obvious, since $T_\omega'(x) > 1$ for $x > 0$ and $x = 0$ is the unique fixed point in $[0, 1/2]$. Since all the maps in $\mathcal{F}$ are uniformly expanding except at 0, there exists $n_0 \in \mathbb{N}$ independent of ω such that $X_n(\omega) \in (0, \varepsilon_0)$ for all $n \ge n_0$. Thus, it is sufficient to prove inequality (4) for any $n \ge n_0$. We now define a sequence $\{Z_n\}_n$ which bounds $X_n(\omega)$ from below and has desired asymptotic. Let $K_0 = [0, \varepsilon_0] \times [\alpha_0, \alpha_1]$ and $C_1 = \max_{(x,\alpha)\in K_0} c_\alpha(1 + f_\alpha(x))$. Set $G(x) = x(1 + C_1 x^{\alpha_0})$. Define $\{Z_n\}_{n\ge n_0}$ as follows: $Z_{n_0} = \min_{\omega\in\Omega} X_{n_0}(\omega)$ and let $Z_n = (G|_{[0,\varepsilon_0]})^{-1}(Z_{n-1})$ for $n > n_0$. Since $G(x) \ge T_{\alpha(\omega)}(x)$ for any $x \in [0, \varepsilon_0]$ and for any $\omega \in \Omega$, one can easily verify by induction that $Z_n \le X_n(\omega)$ for $n \ge n_0$. Finally note that $Z_n \sim n^{-1/\alpha_0}$ [9]. Defining $C_1' = \min_{(x,\alpha)\in K_0} c_\alpha(1 + f_\alpha(x))$, $G'(x) = x(1 + C_1' x^{\alpha_1})$, $Z_{n_0}' = \max_{\omega\in\Omega} X_{n_0}(\omega)$ and $Z_n' = (G'|_{[0,\varepsilon_0]})^{-1}(Z_{n-1}')$ for $n > n_0$ we obtain a sequence $\{Z_n'\}$ such that $X_n(\omega) \le Z_n'$ and $Z_n' \sim n^{-1/\alpha_1}$. This finishes the proof.

Item (2) is proved below. Note that by the choice of n_0 for any $n \ge n_0$ we have

$$X_n(\sigma\omega) = X_{n+1}(\omega)[1 + c_{\alpha(\omega)} X_{n+1}(\omega)^{\alpha(\omega)}(1 + f_{\alpha(\omega)} \circ X_{n+1}(\omega))]. \tag{10}$$

The latter equality together with the standard estimate $(1 + x)^{-a} \leq 1 - ax + \frac{a(a+1)}{2}x^2$ for $x, a > 0$ implies that

$$\frac{1}{X_{n+1}(\omega)^{\alpha_0}} - \frac{1}{X_n(\sigma\omega)^{\alpha_0}} \geq C_1 \alpha_0 X_{n+1}(\omega)^{\alpha(\omega)-\alpha_0} - C_2 X_{n+1}(\omega)^{2\alpha(\omega)-\alpha_0},$$

where, $C_2 = \frac{\alpha_0(\alpha_0+1)}{2} \min_{(\alpha,x) \in K_0} [c_\alpha(1 + f_\alpha(x))]^2$. Hence,

$$\frac{1}{X_n(\omega)^{\alpha_0}} \geq \frac{1}{x_{\alpha(\omega)}^{\alpha_0}} + C_1 \alpha_0 \sum_{k=2}^{n} X_k(\omega)^{\alpha(\sigma^{n-k}\omega)-\alpha_0} - C_2 \sum_{k=2}^{n} X_k(\omega)^{2\alpha(\sigma^{n-k}\omega)-\alpha_0},$$

Notice that we can take C_1 and C_2 are independent of ω. Therefore, by inequality (4) we have

$$\frac{1}{X_n(\omega)^{\alpha_0}} \geq 1 + C_3 \sum_{k=2}^{n} (k^{1/\alpha_0})^{\alpha_0 - \alpha(\sigma^{n-k}\omega)} - C_2 \sum_{k=2}^{n} (k^{1/\alpha_1})^{(-2\alpha(\sigma^{n-k}\omega)+\alpha_0)}, \quad (11)$$

To obtain estimates for the right hand side first we will show the right hand side of the latter inequality on average behaves like $n^{-1} \log n$ as n goes to infinity. We set

$$a_k := (k^{1/\alpha_0})^{\alpha_0 - \alpha(\sigma^{n-k}\omega)}, \quad b_k = (k^{1/\alpha_1})^{-2\alpha(\sigma^{n-k}\omega)+\alpha_0}$$

and

$$S_n = \sum_{k=2}^{n} C_3 a_k - C_2 b_k.$$

Lemma 4 *There exists $C_4 > 0$ such that $\lim_{n \to \infty} \dfrac{\log n}{n} E_{\mathbb{P}}(S_n) = C_4$.*

Proof Applying the above lemma to $E_{\mathbb{P}}(e^{\log a_k})$ with $c = 1$ and $t = \log k^{1/\alpha_1}$ and using the fact $\sum_{k \leq n} \frac{1}{\log k} \sim \frac{n}{\log n}$ we obtain

$$\sum_{k=2}^{n} E_{\mathbb{P}}(a_k) = \frac{\alpha_0}{\alpha_1 - \alpha_0} \sum_{k=2}^{n} \frac{1}{\log k} (1 - k^{-\frac{\alpha_1 - \alpha_0}{\alpha_0}}) = \frac{\alpha_0}{\alpha_1 - \alpha_0} \frac{n}{\log n} + O(n^{1 - \frac{\alpha_1 - \alpha_0}{\alpha_0}} (\log n)^{-1})$$

and hence,

$$\frac{\log n}{n} \sum_{k=2}^{n} E_{\mathbb{P}}(a_k) = \frac{\alpha_0}{\alpha_1 - \alpha_0} + O(n^{-\frac{\alpha_1 - \alpha_0}{\alpha_0}}). \quad (12)$$

Similarly, applying Lemma 3 to $E_{\mathbb{P}}(b_k)$ with $c = 2$ and $t = \log k^{1/\alpha_1}$, we obtain

$$\sum_{k=2}^{n} E_{\mathbb{P}}(b_k) := \frac{\alpha_1}{2(\alpha_1 - \alpha_0)} \sum_{k=2}^{n} \frac{1}{\log k} (k^{-\frac{\alpha_0}{\alpha_1}} - k^{\frac{\alpha_0}{\alpha_1} - 2}) = \frac{\alpha_1}{2(\alpha_1 - \alpha_0)} \frac{n^{1 - \alpha_0/\alpha_1}}{\log n} + o(n).$$

and hence,

$$\lim_{n\to\infty} \frac{\log n}{n} \sum_{k=2}^{n} E_{\mathbb{P}}(b_k) = \lim_{n\to\infty} n^{-\alpha_0/\alpha_1} = 0. \tag{13}$$

Combining (12) and (13) implies

$$\lim_{n\to\infty} \frac{\log n}{n} E_{\mathbb{P}}(S_n) = \lim_{n\to\infty} \frac{\log n}{n} \sum_{k=2}^{n} E_{\mathbb{P}}(C_3 a_k - C_2 b_k) = C_4,$$

where $C_4 = C_3 \alpha_0/(\alpha_1 - \alpha_0)$.

Now we construct a random variable $n_1 : \Omega \to \mathbb{N}$ as in item (2) of Proposition 1. Notice that Lemma 4 implies that there exists N independent of ω such that

$$\frac{C_4}{2} \leq \frac{\log n}{n} E_{\mathbb{P}}(S_n) \leq \frac{3C_4}{2} \tag{14}$$

for all $n \geq N$. On the other hand, by [12, Theorem 1], there exists $C > 0$ such that for every $t > 0$ and $n \in \mathbb{N}$ we have

$$\mathbb{P}\left\{ \frac{\log n}{n} |S_{n+1} - E_{\mathbb{P}}(S_{n+1})| < t \right\} \leq e^{-\frac{Cnt^2}{(\log n)^2}}.$$

Thus, by letting $C_5 = CC_4^2/16$ we obtain

$$\mathbb{P}\left\{ \frac{\log n}{n} S_{n+1} < \frac{C_4}{4} \right\} \leq \mathbb{P}\left\{ \frac{\log n}{n} (S_{n+1} - E_{\mathbb{P}} S_{n+1}) < -\frac{C_4}{4} \right\} \leq e^{-\frac{C_5 n}{(\log n)^2}}. \tag{15}$$

Define

$$n_1(\omega) = \inf\left\{ n \geq N \mid \forall k \geq n, \frac{\log k}{k} S_k \geq \frac{C_4}{4} \right\}.$$

Inequality (15) implies that

$$\mathbb{P}\{n_1(\omega) > n\} \leq \sum_{k=n}^{\infty} e^{-\frac{C_5 k}{(\log k)^2}} \leq C_6 \sum_{k=n}^{\infty} e^{-uk^v} \leq C e^{-un^v}$$

for some $C > 0$, $u > 0$ and $v \in (0, 1)$ which proves inequality (5).

For any $n \geq n_1$ by (11) we have

$$X_n(\omega)^{\alpha_0} \leq \frac{\log n}{n} \frac{4}{C_4}.$$

Hence, for some positive $C > 0$ we have

$$X_n(\omega) \leq C \left(\frac{\log n}{n} \right)^{1/\alpha_0}.$$

This finishes the proof of (6). It remains to prove (7). Recall that there exists n_0 which depends only on ε_0 in (A3) such that (10) holds for all $n \geq n_0$. Thus, recalling that σ preserves $\mathbb{P}$ we have

$$\int m\{R_\omega = n\}d\mathbb{P}(\omega) = \int (X_{n-1}(\sigma\omega) - X_n(\sigma\omega))d\mathbb{P}(\omega) =$$

$$\int_{\{n_1(\omega)>n\}} (X_{n-1}(\sigma\omega) - X_n(\omega))d\mathbb{P}(\omega) + \int_{\{n_1(\omega)\leq n\}} (X_{n-1}(\sigma\omega) - X_n(\omega))d\mathbb{P}(\omega)$$

$$\leq Ce^{-un^\nu} + \int_{\{n_1(\omega)\leq n\}} c_{\alpha(\omega)} X_n(\omega)^{\alpha(\omega)+1}(1 + f_{\alpha(\omega)} \circ X_n(\omega)d\mathbb{P}(\omega)$$

$$\leq Ce^{-un^\nu} + C \int \left(\frac{\log n}{n} \right)^{(\alpha(\omega)+1)/\alpha_0} d\mathbb{P}(\omega) \leq C \left(\frac{\log n}{n} \right)^{(\alpha_0+1)/\alpha_0}.$$

This finishes the proof for all $n \geq n_0$. For $n < n_0$ the assertion follows by increasing the constant C if necessary.

Acknowledgements This research was supported by The Leverhulme Trust through the research grant RPG-2015-346. The author would like to thank Wael Bahsoun for useful discussions during the preparation of the paper.

References

1. Aimino, R., Hu, H., Nicol, M., Török, A., Vaienti, S.: Polynomial loss of memory for maps of the interval with a neutral fixed point. Discrete Contin. Dyn. Syst. **35**(3), 793–806 (2015)
2. Ayyer, A., Liverani, C., Stenlund, M.: Quenched CLT for random toral automorphisms. Discrete Contin. Dyn. Syst. **24**(2), 331–348 (2009)
3. Bahsoun, W., Bose, C.: Mixing rates and limit theorems for random intermittent maps. Nonlinearity **29**(4), 1417–1433 (2016)
4. Bahsoun, W., Bose, C., Duan, Y.: Decay of correlation for random intermittent maps. Nonlinearity **27**(7), 1543–1554 (2014)
5. Bahsoun, W., Bose, C., Ruziboev, M.: Quenched Decay of Correlations for Slowly Mixing Systems. arXiv:1706.04158. Cited 30 Jan 2018
6. Baladi, V., Benedicks, M., Maume-Deschamps, V.: Almost sure rates of mixing for i.i.d. unimodal maps. (English, French summary) Ann. Sci. École Norm. Sup. (4) **35**(1), 77–126 (2002)
7. Buzzi, J.: Exponential decay of correlations for random Lasota-Yorke maps. Comm. Math. Phys. **208**(1), 25–54 (1999)
8. de Melo, W., van Strien, S.: One-Dimensional Dynamics. Springer, Berlin (1993)
9. Gouëzel, S.: Sharp polynomial estimates for the decay of correlations. Israel J. Math. **139**, 29–65 (2004)
10. Haydn, N., Nicol, M., Török, A., Vaienti, S.: Almost sure invariance principle for sequential and non-stationary dynamical systems. Trans. Amer. Math. Soc. **369**(8), 5293–5316 (2017)
11. Haydn, N., Rousseau, J., Yang, F.: Exponential Law for Random Maps on Compact Manifolds. arXiv:1705.05869. Cited 16 May 2017

12. Hoeffding, W.: Probability inequalities for sums of bounded random variables. J. Amer. Stat. Soc. **58**(30), 13–30 (1963)
13. Keller, G.: Generalized bounded variation and applications to piecewise monotonic transformations. Z. Wahr. Verw. Geb. **69**, 461–478 (1985)
14. Kifer, Y.: Limit theorems for random transformations and processes in random environments. Trans. Amer. Math. Soc. **350**(4), 1481–1518 (1998)
15. Leppänen, J., Stenlund, M.: Quasistatic dynamics with intermittency. Math. Phys. Anal. Geom. **19**(2), Art. 8, 23 pp (2016)
16. Liverani, C., Saussol, B., Vaienti, S.: A probabilistic approach to intermittency. Ergod. Theory Dyn. Syst. **19**, 671–685 (1999)
17. Nicol, M., Török, A., Vaienti, S.: Central limit theorems for sequential and random intermittent dynamical systems. Ergod. Theory Dyn. Syst. https://doi.org/10.1017/etds.2016.69
18. Pomeau, Y., Manneville, P.: Intermittent transition to turbulence in dissipative dynamical systems. Comm. Math. Phys. **74**, 189–197 (1980)
19. Saussol, B.: Absolutely continuous invariant measures for multidimensional expanding maps. Israel J. Math. **116**, 223–248 (2000)
20. Young, L.-S.: Recurrence times and rates of mixing. Israel J. Math. **110**, 153–188 (1999)

Conjugations Between Two Critical Circle Maps With Non-integer Exponents

Utkir Safarov

Abstract Let f_1 and f_1 be orientation preserving circle homeomorphisms with single critical point of non-integer order and same irrational rotation numbers. We prove that if the orders of critical points are different then the map h conjugating f_1 and f_2 is a singular function.

Keywords Circle homeomorphisms · Rotation number · Critical point
Conjugation map · Singular function

1 Introduction

In this paper we study a class of circle homeomorphisms with single critical point i.e. the derivative vanishes at the point. The classification of circle homeomorphisms under change of variables is one of the important problems in one-dimensional dynamics. It was started by Poincaré who was motivated by studies in differential equations more than a century ago and has been actively studied ever since.

We identify the unit circle $S^1 = R^1/Z^1$ with the half open interval $[0, 1)$. Let f be a circle homeomorphism that preserves orientation, i.e. $f(x) = F(x)(mod\ 1)$, $x \in S^1 \simeq [0, 1)$, where F is continuous, strictly increasing on R^1 and $F(x + 1) = F(x) + 1$ for any $x \in R$. F is called **lift** of homeomorphism f. The most important arithmetic characteristic of the homeomorphism f of the unit circle S^1 is the rotation number. If f is a circle homeomorphism with lift F, then rotation number $\rho = \rho_f$ is defined by

$$\rho_f = \lim_{n \to \infty} \frac{F^n(x)}{n}(mod\ 1),$$

with F^n the nth iterate of F. The rotation number is rational if and only if f has periodic points. By Denjoy's classical theorem [4], any C^2 circle diffeo-

U. Safarov (✉)
Turin Politechnic University in Tashkent, Kichik halqa yuli, 17,
100095 Tashkent, Uzbekistan
e-mail: safarovua@mail.ru

© Springer Nature Switzerland AG 2018
A. Azamov et al. (eds.), *Differential Equations and Dynamical Systems*,
Springer Proceedings in Mathematics & Statistics 268,
https://doi.org/10.1007/978-3-030-01476-6_12

morphism f with irrational rotation number are conjugate to the linear rotation $f_\rho : x \to x + \rho(mod\ 1)$, that is, there exists an essentially unique homeomorphism φ of the circle with $\varphi \circ f = f_\rho \circ \varphi$. Since the conjugating map φ and the unique f-invariant measure ν_f are related by $\varphi(x) = \nu_f([0, x])$, $x \in S^1$ (see [3]), regularity properties of the conjugating map φ imply corresponding properties of the density of the absolutely continuous invariant measures ν_f. The problem of relating the smoothness of φ to that of f has been studied extensively. In depth results have been found in works ([1, 8, 9, 12, 14, 16]). Note that for sufficiently smooth circle diffeomorphism f with a typical irrational rotation number the conjugacy φ is C^1-homeomorphism. Consequently, the invariant measure ν_f is absolutely continuous w.r.t. Lebesgue measure ℓ on S^1.

A natural extension of circle diffeomorphisms is circle homeomorphisms with critical points.

Definition 1 The point $x_{cr} \in S^1$ is called non-flat critical point of a homeomorphism f with order $d > 1$, if for a some $\delta-$ neighborhood $U_\delta(x_{cr})$ such that $f(x) = \phi(x)|\phi(x)|^{d-1} + f(x_{cr})$ for all $x \in U_\delta(x_{cr})$, where $\phi : U_\delta(x_{cr}) \to \phi(U_\delta(x_{cr}))$ is a C^3 diffeomorphism such that $\phi(x_{cr}) = 0$.

By a **critical circle map** we define an orientation preserving circle homeomorphism with exactly one non-flat critical point.

An important one-parameter family of examples of critical circle maps are the Arnold's maps defined by

$$f_\theta(x) := x + \theta + \frac{1}{2\pi} \sin 2\pi x \ (mod\ 1), \ x \in S^1$$

For every $\theta \in R^1$ the map f_θ is a critical map with critical point 0 of cubic type. Yoccoz in [16] generalized Denjoy's classical result, a critical circle homeomorphism with irrational rotation number is topologically conjugate to an irrational rotation.

The singularity of conjugating map for critical circle homeomorphisms was shown by Graczyk and Swiatek in [7]. They proved that if f is C^3 smooth circle homeomorphism with finitely many critical points of polynomial type and an irrational rotation number of bounded type, then the conjugating map φ is a singular function on S^1 i.e. $\varphi'(x) = 0$ a.e. with respect to the Lebegue measure.

Hence the problem of regularity of the conjugacy between two critical maps with identical irrational rotation number arises naturally. This is called the rigidity problem for critical circle homeomorphisms. For the critical circle maps with odd order (i.e. $d = 2m + 1$, $m \in N$) the rigidity problem is developed by de Faria, de Melo, Yampolsky, Khanin and Teplinsky among others. Initial results connected with rigidity for critical maps were proven by de Faria and de Melo [6]. They proved that any two C^3 critical circle maps with the same order of the critical points (given by odd integer numbers) and with the same irrational rotation number of bounded type are $C^{1+\varepsilon}$, $\varepsilon > 0$ smoothly conjugate to each other. Later D. Khmelev and M. Yampolsky [13] showed that in the analytic case the conjugacy is $C^{1+\alpha}$ smooth near the critical point. A. Avila [2] showed, that there exist f_1 and f_2 analytic homeomorphisms with the same irrational rotation number such that h is not $C^{1+\alpha}$ for any $\alpha > 0$.

Next we formulate the fundamental result of K. Khanin and A. Teplinsky [11].

Theorem 1 *Let f_1 and f_2 be two analytic critical circle maps with the same order of critical points and the same irrational rotation number. Then they are C^1-smoothly conjugate to each other.*

Here we consider the case, when the order of critical points are non-integer and different. Now we formulate our main result.

Theorem 2 *Let f_1 and f_2 be critical circle maps with the same irrational rotation number. Suppose that (i) f_i, $i = 1, 2$ have an unique critical point $x_{cr}^{(i)}$ with order $m_i \in R^1$, $m_i > 2$ and $m_1 \neq m_2$;*

(ii) $f_i(x) = (x - x_{cr}^{(i)})|x - x_{cr}^{(i)}|^{m_i-1} + f_i(x_{cr}^{(i)})$ *for some ω_i-neighborhood $U_{\omega_i}(x_{cr}^{(i)})$ and $f_i \in C^3(S^1 \setminus U_{\omega_i}(x_{cr}^{(i)}))$*

Then the conjugacy h between f_1 and f_2 is a singular function on S^1.

2 Dynamical Partitions, Cross-Ratio Tools

2.1 Dynamical Partition

Let f be an orientation preserving homeomorphism of the circle with lift F and irrational rotation number $\rho = \rho_f$. We denote by $\{a_n, n \in N\}$ the sequence of entries in the continued fraction expansion of ρ, i.e. $\rho = [a_1, a_2, \ldots, a_n, \ldots]$. Denote by $p_n/q_n = [a_1, a_2, \ldots, a_n]$ the convergents of ρ. Their denominators q_n satisfy the recurrence relation, that is $q_{n+1} = a_{n+1}q_n + q_{n-1}$, $n \geq 1$, $q_0 = 1$, $q_1 = a_1$.

For an arbitrary point $x_0 \in S^1$ we define $\Delta_0^{(n)}(x_0)$ the closed interval on S^1 with endpoints x_0 and $x_{q_n} = f^{q_n}(x_0)$. Note that for odd n the point x_{q_n} lies to the left of x_0 and for even n to the right. Denote by $\Delta_i^{(n)}(x_0)$ the iterates of the interval $\Delta_0^{(n)}(x_0)$ under $f:\Delta_i^{(n)}(x_0) := f^i(\Delta_0^{(n)}(x_0))$, $i \geq 1$.

Lemma 1 (see [12]) *Consider an arbitrary point $x_0 \in S^1$. A finite piece $\{x_i, 0 \leq i < q_n + q_{n-1}\}$ of the trajectory of this point divides the circle into the following disjoint (except for the endpoints) intervals: $\Delta_i^{(n-1)}(x_0)$, $0 \leq i < q_n$, $\Delta_j^{(n)}(x_0)$, $0 \leq j < q_{n-1}$.*

We denote the obtained partition by $\xi_n(x_0)$ and call it nth **dynamical partition** of the circle. We now briefly describe the process of transition from $\xi_n(x_0)$ to $\xi_{n+1}(x_0)$. All intervals $\Delta_j^{(n)}(x_0)$, $0 \leq j < q_{n-1}$, are preserved, and each of the intervals $\Delta_i^{(n-1)}(x_0)$ is divided into $a_{n+1} + 1$ sub intervals:

$$\Delta_i^{(n-1)}(x_0) = \Delta_i^{(n+1)}(x_0) \cup \bigcup_{s=0}^{a_{n+1}-1} \Delta_{i+q_{n-1}+sq_n}^{(n)}(x_0).$$

Obviously one has $\xi_1(x_0) \le \xi_2(x_0) \le \ldots \le \xi_n(x_0) \le \ldots$.

Definition 2 Let $K > 1$ be a constant. We call two intervals I_1 and I_2 of S^1 are $K-$**comparable**, if the inequalities $K^{-1}\ell(I_2) \le \ell(I_1) \le K\ell(I_2)$ hold.

Next we formulate the lemma, that is proved in the similar way as in [15].

Let $x_{cr} \in S^1$ be a critical point of homeomorphism f. For any $x_0 \in S^1$, consider the dynamical partition $\xi_n(x_0)$. For definiteness we assume that n is odd. Then $x_{q_n} \prec x_0 \prec x_{q_{n-1}}$. The structure of the dynamical partition implies that $\overline{x}_{cr} = f^{-p}(x_{cr}) \in [x_{q_n}, x_{q_{n-1}}]$, for some p, $0 < p < q_n$. Let I_1 and I_2 be any elements of a dynamical partition $\xi_m(\overline{x}_{cr})$, $m \ge n$ having a common endpoints.

Lemma 2 *Let $f \in C^3(S^1)$ be a critical circle homeomorphism with irrational rotation number. Then there exists a constant $K > 1$ depending only on f such that the intervals I_1 and I_2 are K-comparable.*

It follows from the Lemma 2 that the trajectory of each point is dense in S^1. Hence it follows that there exists conjugation map φ between f and f_ρ, i.e. $\varphi(f(x)) = f_\rho(\varphi(x))$ for any $x \in S^1$.

We assume that $\Delta^{(m+k)}$ is element of partitioning $\xi_{m+k}(\overline{x}_{cr})$, while $\Delta^{(m)}$ is an element of partitioning $\xi_m(\overline{x}_{cr})$ that contains $\Delta^{(m+k)}$.

Lemma 3 (see [10], Lemma 2, p.183) *There exist constants $\lambda_1(f) < \lambda_2(f) < 1$ such that*

$$\ell(\Delta^{(m+k)}) \le const\lambda_2^k(f)\ell(\Delta^{(m)}), \ \ \ell(\Delta_0^{(m)}) \ge const\lambda_1^m(f)$$

2.1.1 Cross-Ratio Tools

In the proof of our main theorem the tool of cross-ratio plays a key role.

Definition 3 The **cross-ratio** of four points (z_1, z_2, z_3, z_4), $z_1 < z_2 < z_3 < z_4$ is the number

$$Cr(z_1, z_2, z_3, z_4) = \frac{(z_2 - z_1)(z_4 - z_3)}{(z_3 - z_1)(z_4 - z_2)}.$$

Definition 4 Given four real numbers (z_1, z_2, z_3, z_4) with $z_1 < z_2 < z_3 < z_4$ and a strictly increasing function $F : R^1 \to R^1$. The distortion of their cross-ratio under F is given by

$$Dist(z_1, z_2, z_3, z_4; F) = \frac{Cr(F(z_1), F(z_2), F(z_3), F(z_4))}{Cr(z_1, z_2, z_3, z_4)}.$$

For $m \ge 3$ and $z_i \in S^1$, $1 \le i \le m$, suppose that $z_1 \prec z_2 \prec \cdots \prec z_m \prec z_1$ (in the sense of the ordering on the circle). Then we set $\hat{z}_1 := z_1$ and

$$\hat{z}_i := \begin{cases} z_i, & \text{if } z_1 < z_i < 1, \\ 1 + z_i, & \text{if } 0 < z_i < z_1. \end{cases}$$

for $2 \leq i \leq m$.

Obviously, $\hat{z}_1 < \hat{z}_2 < \cdots < \hat{z}_m$. The vector $(\hat{z}_1, \hat{z}_2, \ldots, \hat{z}_m)$ is called the lifted vector of $(z_1, z_2, \ldots, z_m) \in (S^1)^m$.

Let f be a circle homeomorphism with lift F. We define the cross-ratio distortion of (z_1, z_2, z_3, z_4), $z_1 \prec z_2 \prec z_3 \prec z_4 \prec z_1$ with respect to f by $Dist(z_1, z_2, z_3, z_4; f) = Dist(\hat{z}_1, \hat{z}_2, \hat{z}_3, \hat{z}_4; F)$, where $(\hat{z}_1, \hat{z}_2, \hat{z}_3, \hat{z}_4)$ is the lifted vector of (z_1, z_2, z_3, z_4). We need the following lemma.

Lemma 4 ([5]) *Let* $z_i \in S^1$, $i = 1, 2, 3, 4$, $z_1 \prec z_2 \prec z_3 \prec z_4$. *Consider a circle homeomorphism* f_1 *with* $f_1 \in C^{2+\varepsilon}([z_1, z_4])$, $\varepsilon > 0$, *and* $f_1'(x) \geq const > 0$ *for* $x \in [z_1, z_4]$. *Then there is a positive constant* $C_1 = C_1(f)$ *such that*

$$| Dist(z_1, z_2, z_3, z_4; f_1) - 1 | \leq C_1 |\hat{z}_4 - \hat{z}_1|^{1+\varepsilon},$$

where $(\hat{z}_1, \hat{z}_2, \hat{z}_3, \hat{z}_4)$ *is the lifted vector of* (z_1, z_2, z_3, z_4).

We now consider the case when the interval $[z_1, z_4]$ contains a critical point $x_{cr}^{(1)}$ of the homeomorphism f_1. More precisely, suppose that $z_2 = x_{cr}^{(1)}$. We define numbers α, β, γ, ξ and η as follows:

$$\alpha := \hat{z}_2 - \hat{z}_1, \ \ \beta := \hat{z}_3 - \hat{z}_2, \ \ \gamma := \hat{z}_4 - \hat{z}_3, \ \ \xi := \frac{\beta}{\alpha}, \ \ \eta := \frac{\beta}{\gamma},$$

where $(\hat{z}_1, \hat{z}_2, \hat{z}_3, \hat{z}_4)$ is the lifted vector of (z_1, z_2, z_3, z_4).

Thus we need the following lemma.

Lemma 5 *Suppose that the homeomorphism* f_1 *with lift* F_1 *satisfies the conditions of Theorem 1. Then for any* $z_i \in U_\delta(x_{cr}^{(1)})$, $i = \overline{1, 4}$, $z_1 \prec z_2 = x_{cr}^{(1)} \prec z_3 \prec z_4$ *one has*

$$|Dist(z_1, z_2, z_3, z_4; f_1) = \frac{1 + \xi}{1 + \xi^{m_1}} \cdot \frac{(1 + \eta)^{m_1} - \eta^{m_1}}{(1 + \eta)^{m_1}} \cdot (1 + \eta).$$

Proof By the assumption of the lemma $z_2 = x_{cr}$. We write $Cr(f_1(z_1), f_1(z_2), f_1(z_3), f_1(z_4))$ as follows

$$Cr(f_1(z_1), f_1(z_2), f_1(z_3), f_1(z_4)) = \frac{(F_1(\hat{z}_2) - F_1(\hat{z}_1))(F_1(\hat{z}_4) - F_1(\hat{z}_3))}{(F_1(\hat{z}_3) - F_1(\hat{z}_1))(F_1(\hat{z}_4) - F_1(\hat{z}_2))}, \quad (1)$$

where $(\hat{z}_1, \hat{z}_2, \hat{z}_3, \hat{z}_4)$ is the lifted vector of (z_1, z_2, z_3, z_4). Since $F_1(x) = (x - \hat{x}_{cr}^{(1)})|x - \hat{x}_{cr}^{(1)}|^{m_1-1} + F_1(\hat{x}_{cr}^{(1)})$ on the interval $U_{\omega_1}(\hat{x}_{cr}^{(1)})$ and from (1) we have

$$Cr(f_1(z_1), f_1(z_2), f_1(z_3), f_1(z_4)) = \frac{\alpha^{m_1}}{\alpha^{m_1} + \beta^{m_1}} \cdot \frac{(\alpha + \beta)^{m_1} - \beta^{m_1}}{(\alpha + \beta)^{m_1}}.$$

Using this we get

$$Dist(z_1, z_2, z_3, z_4; f_1) = \frac{\frac{1}{1+\xi^{m_1}}}{\frac{1}{1+\xi}} \cdot \frac{\frac{(1+\eta)^{m_1}-\eta^{m_1}}{(1+\eta)^{m_1}}}{\frac{1}{1+\eta}}.$$

From the last relation it follows

$$Dist(z_1, z_2, z_3, z_4; f_1) = \frac{1+\xi}{1+\xi^{m_1}} \cdot \frac{(1+\eta)^{m_1}-\eta^{m_1}}{(1+\eta)^{m_1}} \cdot (1+\eta).$$

Thus Lemma 5 is proved.

Next suppose the interval $[z_1, z_4]$ is a subset of the interval $U_{\omega_1}(x_{cr}^{(1)})$ but does not contain a critical point $x_{cr}^{(1)}$ of the homeomorphism f_1. Let $d = \min_{1 \leq s \leq 4} \ell([z_s, x_{cr}^{(1)}])$.

Lemma 6 *Let f_1 be a circle homeomorphism satisfying conditions of Theorem 1. Suppose that $d > \alpha + \beta + \gamma$. Then the following equality holds*

$$Dist(z_1, z_2, z_3, z_4; f_1) = 1 + O\left(\left(\frac{\alpha+\beta+\gamma}{d}\right)^2\right).$$

Proof Let f_1 be satisfying the conditions of Theorem 1. For definiteness, we may assume that $x_{cr}^{(1)} \prec z_1 \prec z_2 \prec z_3 \prec z_4$. By definition of cross-ratio implies that

$$Cr(f_1(z_1), f_1(z_2), f_1(z_3), f_1(z_4)) = \frac{(\alpha+d)^{m_1} - d^{m_1}}{(\alpha+\beta+d)^{m_1} - d^{m_1}} \times$$

$$\times \frac{(\alpha+\beta+\gamma+d)^{m_1} - (\alpha+\beta+d)^{m_1}}{(\alpha+\beta+\gamma+d)^{m_1} - (\alpha+d)^{m_1}} = \frac{1 - (1+\frac{\alpha}{d})^{m_1}}{1 - (1+\frac{\alpha+\beta}{d})^{m_1}} \cdot \frac{(1+\frac{\alpha+\beta}{d})^{m_1} - (1+\frac{\alpha+\beta+\gamma}{d})^{m_1}}{(1+\frac{\alpha}{d})^{m_1} - (1+\frac{\alpha+\beta+\gamma}{d})^{m_1}}.$$

The following equalities are easy to check:

$$\left(1+\frac{\alpha}{d}\right)^{m_1} = 1 + m_1\frac{\alpha}{d} + \frac{m_1(m_1-1)}{2}\left(\frac{\alpha}{d}\right)^2 + O\left(\left(\frac{\alpha}{d}\right)^3\right),$$

$$\left(1+\frac{\alpha+\beta}{d}\right)^{m_1} = 1 + m_1\frac{\alpha+\beta}{d} + \frac{m_1(m_1-1)}{2}\left(\frac{\alpha+\beta}{d}\right)^2 + O\left(\left(\frac{\alpha+\beta}{d}\right)^3\right),$$

$$\left(1+\frac{\alpha+\beta+\gamma}{d}\right)^{m_1} = 1 + m_1\frac{\alpha+\beta+\gamma}{d} + \frac{m_1(m_1-1)}{2}\left(\frac{\alpha+\beta+\gamma}{d}\right)^2 + O\left(\left(\frac{\alpha+\beta+\gamma}{d}\right)^3\right).$$

Using these relations we obtain that

$$Dist(z_1, z_2, z_3, z_4; f_1) = \frac{1 + \frac{m_1-1}{2}\frac{\alpha}{d} + O\left(\left(\frac{\alpha}{d}\right)^2\right)}{1 + \frac{m_1-1}{2}\frac{\alpha+\beta}{d} + O\left(\left(\frac{\alpha+\beta}{d}\right)^2\right)} \cdot \frac{1 + \frac{m_1-1}{2}\frac{\alpha}{d} + O\left(\left(\frac{\alpha}{d}\right)^2\right)}{1 + \frac{m_1-1}{2}\frac{\alpha+\beta}{d} + O\left(\left(\frac{\alpha+\beta}{d}\right)^2\right)} =$$

$$= \left(1 + \frac{m_1-1}{2}\frac{\alpha}{d} + O\left(\left(\frac{\alpha}{d}\right)^2\right)\right) \cdot \left(1 - \frac{m_1-1}{2}\frac{\alpha+\beta}{d} + O\left(\left(\frac{\alpha+\beta}{d}\right)^2\right)\right) \cdot \left(1 + \frac{m_1-1}{2}\frac{\alpha}{d} + O\left(\left(\frac{\alpha}{d}\right)^2\right)\right) \cdot$$

$$\cdot \left(1 - \frac{m_1-1}{2}\frac{\alpha+\beta}{d} + O\left(\left(\frac{\alpha+\beta}{d}\right)^2\right)\right) = \left(1 - \frac{m_1-1}{2} \cdot \frac{\beta}{d} + O\left(\left(\frac{\alpha+\beta+\gamma}{d}\right)^2\right)\right) \cdot$$

$$\cdot \left(1 + \frac{m_1-1}{2} \cdot \frac{\beta}{d} + O\left(\left(\frac{\alpha+\beta+\gamma}{d}\right)^2\right)\right) = 1 + O\left(\left(\frac{\alpha+\beta+\gamma}{d}\right)^2\right).$$

Lemma 6 is proved.

3 Proof of Theorem 1

In order to prove Theorem 1 we need several lemmas which we formulate next. Their proofs will be given later. We consider two copies of the unit circle S^1. The homeomorphism f_1 acts on the first circle and f_2 acts on the second one. Assume that f_i, $i = 1, 2$ satisfies the conditions of Theorem 1.

Let φ_1 and φ_2 be conjugations of f_1 and f_2 to linear rotation f_ρ, i.e. $\varphi_1 \circ f_1 = f_\rho \circ \varphi_1$ and $\varphi_2 \circ f_2 = f_\rho \circ \varphi_2$. It is easy to check that the homeomorphisms f_1 and f_2 are conjugated by $h = \varphi_2 \circ \varphi_1^{-1}$, i.e. $h \circ f_1(x) = f_2 \circ h(x)$, $\forall x \in S^1$. Recall that every φ_i, $i = 1, 2$ is unique up to an additional constant. This gives us a possibility to choose h with initial condition $h(x_{cr}^{(1)}) = x_{cr}^{(2)}$.

Notice the conjugation $h(x)$ is continuous function on S^1. It suffices to show that $h'(x) = 0$ for almost all x with respect to the Lebesgue measure. The derivative $h'(x) = 0$ exists for almost all x with respect to the Lebesgue measure because the function h is monotonic. Let us show that $h'(x) = 0$ at all points where the derivative is defined.

Lemma 7 (see [5]) *Assume, that the conjugating homeomorphism $h(x)$ has a positive derivative $h'(x_0) = p_0$ at some point $x_0 \in S^1$, and that the following conditions hold for the points $z_i \in S^1$, $i = 1, .., 4$, with $z_1 \prec z_2 \prec z_3 \prec z_4$, and some constant $R_1 > 1$:*

(a) the intervals $[z_1, z_2]$, $[z_2, z_3]$, $[z_3, z_4]$ are pairwise R_1-comparable;
(b) $\max_{1 \le i \le 4} \ell([z_i, x_0]) \le R_1 \ell([z_1, z_2])$.

Then for any $\varepsilon > 0$ there exists $\delta = \delta(\varepsilon) > 0$ such that

$$|Dist(z_1, z_2, z_3, z_4; h) - 1| \leq C_2 \varepsilon, \qquad (2)$$

if $z_i \in (x_0 - \delta, \ x_0 + \delta)$ for all $i = 1, 2, 3, 4$, where the constant $C_2 > 0$ depends only on R_1, ω_0 and not on ε.

Suppose that $h'(x_0) = p_0$, where $x_0 \in S^1$. Let $\xi_n(x_0)$ be its nth dynamical partition. Put $t_0 := h(x_0)$ and consider the dynamical partition $\tau_n(t_0)$ of t_0 on the second circle determined by the homeomorphism f_2, i.e.

$$\tau_n(t_0) = \{I_i^{(n-1)}(t_0), \ 0 \leq i \leq q_n - 1\} \cup \{I_j^{(n)}(t_0), \ 0 \leq j \leq q_{n-1} - 1\}$$

with $I_0^{(n)}(t_0)$ the closed interval with endpoints t_0 and $f_2^{q_n}(t_0)$. Choose an odd natural number $n_1 = n(f_1, f_2)$ such that the n_1th renormalization neighborhoods $[x_{q_{n_1}}, x_{q_{n_1-1}}]$ and $[t_{q_{n_1}}, t_{q_{n_1-1}}]$ do not contain critical point of f_1 and f_2 respectively. Since the identical rotation number ρ of f_1 and f_2 is irrational, the order of the points on the orbit $\{f_1^k(x_0), \ k \in Z\}$ on the first circle will be precisely the same as the one for the orbit $\{f_2^k(t_0), \ k \in Z\}$ on the second one. This together with the relation $h(f_1(x)) = f_2(h(x))$ for $x \in S^1$ implies that

$$h(\Delta_i^{(n_1-1)}) = I_i^{(n_1-1)}, \ 0 \leq i \leq q_{n_1} - 1, \ h(\Delta_j^{(n_1)}) = I_j^{(n_1)}, \ 0 \leq j \leq q_{n_1-1} - 1. \qquad (3)$$

The structure of the dynamical partitions implies that $\overline{x}_{cr}^{(1)}(n_1) = f_1^{-l}(x_{cr}^{(1)}) \in [x_{q_{n_1}}, x_{q_{n_1-1}}]$, where $l \in (0, q_{n_1}-1)$ if $\overline{x}_{cr}^{(1)}(n_1) \in [x_{q_{n_1}}, x_0]$, and $l \in (0, q_{n_1})$ if $\overline{x}_{cr}^{(1)}(n_1) \in [x_0, x_{q_{n_1-1}}]$. Since h conjugation between f_1 and f_2, we get

$$f_2^l(h(\overline{x}_{cr}^{(1)})) = f_2^{l-1}(f_2(h(\overline{x}_{cr}^{(1)}))) = f_2^{l-1}(h(f_1(\overline{x}_{cr}^{(1)}))) = \cdots = h(f_1^l(\overline{x}_{cr}^{(1)})) = h(x_{cr}^{(1)}) = x_{cr}^{(2)}.$$

Hence $\overline{x}_{cr}^{(2)}(n_1) = f_2^{-l}(x_{cr}^{(2)}) \in [t_{q_{n_1}}, t_{q_{n_1-1}}]$. The points $\overline{x}_{cr}^{(1)}(n_1)$ and $\overline{x}_{cr}^{(2)}(n_1)$ are called the q_{n_1}-**preimages** of the critical points $x_{cr}^{(1)}$ and $x_{cr}^{(2)}$, respectively.

Next we introduce the concept of a "regular" cover of the critical point. Let $z_i \in S^1$, $i = \overline{1,4}$, $z_1 \prec z_2 \prec z_3 \prec z_4 \prec z_1$. Define for each j, $0 < j < q_n$

$$\xi_{f_1}(j) = \frac{\ell([f_1^j(z_2), f_1^j(z_3)])}{\ell([f_1^j(z_1), f_1^j(z_2)])}, \quad \eta_{f_1}(j) = \frac{\ell([f_1^j(z_2), f_1^j(z_3)])}{\ell([f_1^j(z_3), f_1^j(z_4)])}.$$

Definition 5 Let $M > 1$, $\zeta \in (0, 1)$, $\delta > 0$ be constant numbers, n - a positive integer and $x_0 \in S^1$. We say that a triple of intervals $([z_1, z_2], [z_2, z_3], [z_3, z_4])$, $z_i \in S^1$, $i = 1, 2, 3, 4$, covers the critical point of $x_{cr}^{(1)}$ "$(M, \zeta, \theta, \delta; x_0)$-regularly", if the following conditions hold:

(1) $[z_1, z_4] \subset (x_0 - \delta, x_0 + \delta)$, and the system of intervals $\{f_1^j([z_1, z_4]), \ 0 \leq j \leq q_n - 1\}$ cover critical point $x_{cr}^{(1)}$ only once;

(2) $z_2 = f_1^{-l}(x_{cr}^{(1)})$ for some l, $0 < l < q_n$;

(3) $\xi_{f_1}(l) < \zeta$ and $\eta_{f_1}(l) \geq M$.

Denote

$$L = \min\{m_1, m_2, |m_1 - m_2|\}.$$

Lemma 8 *Suppose that the homeomorphisms f_i, $i = 1, 2$ satisfy the conditions of Theorem 1. Then for any $x_0 \in S^1$ and $\delta > 0$ there exist constant $M_0 > 1$ and $\zeta_0 \in (0, 1)$, such that for all triples of intervals $[z_s, z_{s+1}] \subset (x_0 - \delta, x_0 + \delta)$, $s = 1, 2, 3$, and $[h(z_s), h(z_{s+1})]$, $s = 1, 2, 3$, covering the critical points $x_{cr}^{(1)}$ and $x_{cr}^{(2)}$ regularly with constants M_0 and ζ_0 the following inequalities hold:*

$$\left| \frac{1 + \xi_{f_1}(l)}{1 + \xi_{f_1}^{m_1}(l)} \cdot \frac{(1 + \eta_{f_1}(l))^{m_1} - \eta_{f_1}^{m_1}(l)}{(1 + \eta_{f_1}(l))^{m_1}} \cdot (1 + \eta_{f_1}(l)) - m_1 \right| < \frac{L}{16},$$

$$\left| \frac{1 + \xi_{f_2}(l)}{1 + \xi_{f_2}^{m_1}(l)} \cdot \frac{(1 + \eta_{f_2}(l))^{m_1} - \eta_{f_2}^{m_1}(l)}{(1 + \eta_{f_2}(l))^{m_1}} \cdot (1 + \eta_{f_2}(l)) - m_2 \right| < \frac{L}{16},$$

where m_1 and m_2 are orders of critical points $x_{cr}^{(1)}$ and $x_{cr}^{(2)}$ respectively.

Assume that the homeomorphism f_1 satisfies the conditions of Theorem 1. Let $\xi_n(x_{cr}^{(1)})$ be a dynamical partition of the circle by f_1. We take a natural number r, such that $\Delta_0^{(r)}(x_{cr}^{(1)}) \cup \Delta_0^{(r-1)}(x_{cr}^{(1)}) \subset U_{\omega_1}(x_{cr}^{(1)})$. Suppose that $h'(x_0) = p_0 > 0$ for some $x_0 \in S^1$. Consider the dynamical partition $\xi_n(x_0)$ of the point x_0 under f_1. Suppose that $n > r$ an odd natural number. Let $\overline{x}_{cr}^{(1)} = f^{-l}(x_{cr}^{(1)}) \in [x_{q_n}, x_{q_{n-1}}]$.

Let $\{\xi_{n+k}(\overline{x}_{cr}^{(1)})\}_{k=0}^{\infty}$ be a sequence of dynamical partitions of the point $\overline{x}_{cr}$. We define the points z_i, $i = 1, 2, 3, 4$ as follows

$$z_1 = f^{q_{n+k_0}}(\overline{x}_{cr}^{(1)}), \quad z_2 = \overline{x}_{cr}^{(1)}, \quad z_3 = f^{q_{n+k_0+k_1}}(\overline{x}_{cr}^{(1)}), \quad z_4 = f^{q_{n+k_0+k_1}+q_{n+k_2}}(\overline{x}_{cr}^{(1)}).$$

Lemma 9 *Suppose that the homeomorphisms f_1 and f_2 satisfies the conditions of Theorem 1. Let $h'(x_0) = p_0 > 0$ for some $x_0 \in S^1$, $\delta \in (0, 1)$ and $k_0 \in N$. Then there exist natural numbers k_1, k_2 such that for sufficiently large n, the triple of intervals $[z_s, z_{s+1}] \subset (x_0 - \delta, x_0 + \delta)$, $s = 1, 2, 3$ satisfies the following properties:*

(1) the intervals $\{[f_1^j(z_1), f_1^j(z_4)], 0 \leq j \leq q_n\}$ cover each point at most once;

(2) the intervals $[z_s, z_{s+1}]$ and $[f_1^{q_n}(z_s), f_1^{q_n}(z_{s+1})]$ $s = 1, 2, 3$ satisfy conditions (a) and (b) of Lemma 7 with some constant $R_1 > 1$ depending on k_0, k_1, k_2;

(3) the triples of intervals $([z_s, z_{s+1}], s = 1, 2, 3)$ and $([h(z_s), h(z_{s+1})], s = 1, 2, 3)$ cover the critical points $x_{cr}^{(1)}, x_{cr}^{(2)}$, "$(M_0, \zeta_0, \delta; x_0)$-regularly" and "$(M_0, \zeta_0, \delta; h(x_0))$-regularly", respectively.

Lemma 10 *Suppose the circle homeomorphisms f_1 and f_2 satisfy the conditions of Theorem 1. Then there exists natural number k_0 such that for intervals $[z_s, z_{s+1}]$, $s = 1, 2, 3$ satisfying conditions (1)-(3) of Lemma 9, and for sufficiently large n the following inequality holds*

$$\left| \frac{Dist(z_1, z_2, z_3, z_4; f_1^{q_n})}{Dist(h(z_1), h(z_2), h(z_3), h(z_4); f_2^{q_n})} - 1 \right| \geq R_2 > 0, \tag{4}$$

where the constant R_2 depends only on f_1 and f_2.

Proof of Theorem 1 Let f_1 and f_2 be circle homeomorphisms satisfying the conditions of Theorem 1. The lift $H(x)$ of the conjugating map $h(x)$ is a continuous and monotone increasing function on R^1. Hence $H(x)$ has a finite derivative $H'(x)$ for almost all x with respect to Lebesgue measure. We claim that $h'(x) = 0$ at all points x where the finite derivative exists. Suppose $h'(x_0) > 0$ for some point $x_0 \in S^1$. Fix $\varepsilon > 0$. We take a triple of intervals $[z_s, z_{s+1}] \subset (x_0 - \delta, \ x_0 + \delta)$, $s = 1, 2, 3$, satisfying the conditions of Lemma 10.

Using the assertion of Lemma 7 we obtain

$$|Dist(z_1, z_2, z_3, z_4; h) - 1| \leq C_3\varepsilon, \tag{5}$$

$$|Dist(f_1^{q_n}(z_1), f_1^{q_n}(z_2), f_1^{q_n}(z_3), f_1^{q_n}(z_4); h) - 1| \leq C_3\varepsilon. \tag{6}$$

Hence

$$\left| \frac{Dist(z_1, z_2, z_3, z_4; h)}{Dist(f_1^{q_n}(z_1), f_1^{q_n}(z_2), f_1^{q_n}(z_3), f_1^{q_n}(z_4); h)} - 1 \right| \leq C_4\varepsilon, \tag{7}$$

where the constant $C_4 > 0$ does not depend on ε and n.

Since h is conjugating f_1 and f_2 we can readily see that

$$Cr(h(f_1^{q_n}(z_1)), h(f_1^{q_n}(z_2)), h(f_1^{q_n}(z_3)), h(f_1^{q_n}(z_4))) =$$

$$= Cr(f_2^{q_n}(h(z_1)), f_2^{q_n}(h(z_2)), f_2^{q_n}(h(z_3)), f_2^{q_n}(h(z_4))).$$

Hence we obtain

$$\frac{Dist(f_1^{q_n}(z_1), f_1^{q_n}(z_2), f_1^{q_n}(z_3), f_1^{q_n}(z_4); h)}{Dist(z_1, z_2, z_3, z_4; h)} =$$

$$= \frac{Cr(h(f_1^{q_n}(z_1)), h(f_1^{q_n}(z_2)), h(f_1^{q_n}(z_3)), h(f_1^{q_n}(z_4)))}{Cr(f_1^{q_n}(z_1), f_1^{q_n}(z_2), f_1^{q_n}(z_3), f_1^{q_n}(z_4))} \times$$

$$\times \frac{Cr(z_1, z_2, z_3, z_4)}{Cr(h(z_1), h(z_2), h(z_3), h(z_4))} = \frac{Cr(f_2^{q_n}(h(z_1)), f_2^{q_n}(h(z_2)), f_2^{q_n}(h(z_3)), f_2^{q_n}(h(z_4)))}{Cr(h(z_1), h(z_2), h(z_3), h(z_4))} :$$

$$: \frac{Cr(f_1^{q_n}(z_1), f_1^{q_n}(z_2), f_1^{q_n}(z_3), f_1^{q_n}(z_4))}{Cr(z_1, z_2, z_3, z_4)} = \frac{Dist(h(z_1), h(z_2), h(z_3), h(z_4); f_2^{q_n})}{Dist(z_1, z_2, z_3, z_4; f_1^{q_n})}.$$

This, together with (7) obviously implies that

$$\left| \frac{Dist(z_1, z_2, z_3, z_4; f_1^{q_n})}{Dist(h(z_1), h(z_2), h(z_3), h(z_4); f_2^{q_n})} - 1 \right| \le C_5 \varepsilon,$$

where the constant $C_5 > 0$ does not depend on ε and n. This contradicts equation (4). Therefore Theorem 1 is completely proved.

4   The Proofs of Lemmas 8–10

Proof of Lemma 8 Denote

$$\psi_1(\xi_{f_1}(l)) = \frac{1 + \xi_{f_1}(l)}{1 + \xi_{f_1}^{m_1}(l)},$$

and

$$\psi_2(\eta_{f_1}(l)) = \frac{(1 + \eta_{f_1}(l))^{m_1} - \eta_{f_1}^{m_1}(l)}{(1 + \eta_{f_1}(l))^{m_1}} \cdot (1 + \eta_{f_1}(l)).$$

It is easy to check that for $\eta_{f_1}(l) > 0$ the function $\psi_2(\eta_{f_1}(l))$ is monotone increasing and $1 < \psi_2(\eta_{f_1}(l)) < m_1$. Obviously

$$\lim_{\xi_{f_1}(l) \to 0} \psi_1(\xi_{f_1}(l)) = 1, \quad \lim_{\eta_{f_1}(l) \to \infty} \psi_2(\eta_{f_1}(l)) = m_1.$$

Taking these remarks into account and using the explicit form of the functions $\psi_1(\xi_{f_1}(l))$ and $\psi_2(\eta_{f_1}(l))$ we can now estimate $| \psi_1 \cdot \psi_2 - m_1) |$. Firstly, we estimate ψ_2 for large value of $\eta_{f_1}(l)$. Using the explicit form of the function $\psi_2(\eta_{f_1}(l))$, we see that the inequality

$$|\psi_2 - m_1| = O\left(\frac{1}{\eta_{f_1}(l)}\right) \le R_3 \left(\frac{1}{\eta_{f_1}(l)}\right), \tag{8}$$

where the constant $R_3 > 0$ depends only on f_1. If we choose $\eta_{f_1}(l)$ satisfying the inequality $R_3 \left(\frac{1}{\eta_{f_1}(l)}\right) < \frac{L}{32}$, then

$$|\psi_2(\eta_{f_1}(l)) - m_1| < \frac{L}{32},$$

for $\eta_{f_1}(l) > \frac{32 R_3}{L}$.

We next estimate $|\psi_1 - 1|$ for small value of $\xi_{f_1}(l)$. Using the explicit form of the function $\psi_1(\xi_{f_1}(l))$, we see that $|\psi_1(\xi_{f_1}(l)) - 1| = O(\xi_{f_1}(l)) \le R_4 \xi_{f_1}(l)$. It follows from this together with (8) that $|\psi_1 \cdot \psi_2 - m_1| \le |\psi_2 - m_1| + |\psi_2| \cdot |\psi_1 - 1| \le \frac{L}{32} + m_1 R_4 \xi_{f_1}(l)$. If we take

$$\zeta_1 := \min\{\frac{L}{32m_1R_5}, 1\}, \quad M_1 := \max\{\frac{32R_5}{L}, 1\},$$

where $R_5 = \max\{R_3, R_4\}$, then for all $\xi_{f_1}(l) < \zeta_1$ and $\eta_{f_1}(l) > M_1$ the following inequality holds

$$|\psi_1 \cdot \psi_2 - m_1| \le \frac{L}{16}.$$

Similarly it can be shown that with

$$\zeta_2 := \min\{\frac{L}{32m_2R_6}, 1\}, \quad M_2 := \max\{\frac{32R_6}{L}, 1\}, \tag{9}$$

and $\xi_{f_2}(l) < \zeta_2$ and $\eta_{f_2}(l) > M_2$, the second assertion of Lemma 8 holds. In (9) the constants $R_6 > 0$ depends only on f_2. Finally, if we set $\zeta_0 := \min\{\zeta_1, \zeta_2\}$ and $M_0 := \max\{M_1, M_2\}$, then Lemma 8 holds for $\xi_{f_1}(l), \xi_{f_2}(l) \in [0, \zeta_0)$ and $\eta_{f_1}(l), \eta_{f_2}(l) \ge M_0$. Lemma 8 is proved.

Proof of Lemma 9 Firstly, we prove the third assertion of the lemma. By the construction of the points z_i, $i = 1, 2, 3, 4$, it implies that the intervals $[z_s, z_{s+1}]$ and $[h(z_s), h(z_{s+1})]$, $s = 1, 2, 3$ satisfy the 1) and 2) conditions of definition of "regularly" covering. We consider dynamical partition $\xi_n(x_{cr}^{(1)})$. According to Lemma 2 the intervals $\Delta_0^{(n)}(x_{cr}^{(1)})$ and $\Delta_0^{(n-1)}(x_{cr}^{(1)})$ are K-comparable, i. e. there exist constant $K > 1$ such that $K^{-1}\ell(\Delta_0^{(n-1)}(x_{cr}^{(1)})) \le \ell(\Delta_0^{(n)}(x_{cr}^{(1)})) \le K\ell(\Delta_0^{(n-1)}(x_{cr}^{(1)}))$. Thus it follows that there exists $k_1^{(1)} \in N$ such that the following inequality holds

$$\frac{\ell([x_{cr}^{(1)}, f_1^{q_{n+k_0+k_1^{(1)}}}(x_{cr}^{(1)})])}{\ell([f_1^{q_{n+k_0}}(x_{cr}^{(1)}), x_{cr}^{(1)}])} < \zeta_0. \tag{10}$$

Indeed, it is clear that

$$\frac{\ell(\Delta_0^{(q_{n+k_0}+3)}(x_{cr}^{(1)}))}{\ell(\Delta_0^{(q_{n+k_0}+1)}(x_{cr}^{(1)}))} = \frac{1}{\frac{\ell(\Delta_0^{(q_{n+k_0}+1)}(x_{cr}^{(1)}))}{\ell(\Delta_0^{(q_{n+k_0}+3)}(x_{cr}^{(1)}))}} \le \frac{1}{1 + \frac{1}{K}} = \frac{K}{K+1}.$$

Hence $\ell(\Delta_0^{(q_{n+k_0}+3)}(x_{cr}^{(1)})) \le \frac{K}{K+1}\ell(\Delta_0^{(q_{n+k_0}+1)}(x_{cr}^{(1)}))$. Using the last inequality we obtain that for any k

$$\ell(\Delta_0^{(q_{n+k_0}+k)}(x_{cr}^{(1)})) \le (\frac{K}{K+1})^k \ell(\Delta_0^{(q_{n+k_0}+1)}(x_{cr}^{(1)})).$$

Since $\Delta_0^{(q_{n+k_0}+1)}(x_{cr}^{(1)}))$ and $\Delta_0^{(q_{n+k_0})}(x_{cr}^{(1)}))$ are K-comparable, there exists a $k_1^{(1)} \in N$ such that the inequality (10) is true. Similarly, we can show that there exists a $k_2^{(1)} \in N$ such that the following inequality holds

$$\frac{\ell([x_{cr}^{(1)}, f_1^{q_{n+k_0+k_1}}(x_{cr}^{(1)})])}{\ell([f_1^{q_{n+k_0+k_1}}(x_{cr}^{(1)}), f_1^{q_{n+k_0+k_1}^{(1)}+q_{n+k_2}^{(1)}}(x_{cr}^{(1)})])} > M_0.$$

Similarly, it can be shown that with natural numbers $k_1^{(2)}$ and $k_1^{(2)}$ the inequalities

$$\frac{\ell([x_{cr}^{(2)}, f_2^{q_{n+k_0+k_1}^{(2)}}(x_{cr}^{(2)})])}{\ell([f_2^{q_{n+k_0}}(x_{cr}^{(2)}), x_{cr}^{(2)}])} < \zeta_0, \qquad \frac{\ell([x_{cr}^{(2)}, f_2^{q_{n+k_0+k_1}^{(2)}}(x_{cr}^{(2)})])}{\ell([f_2^{q_{n+k_0+k_1}^{(2)}}(x_{cr}^{(2)}), f_2^{q_{n+k_0+k_1}+q_{n+k_2}^{(2)}}(x_{cr}^{(2)})])} > M_0$$

hold. If we take $k_1 = \max\{k_1^{(1)}, k_1^{(2)}\}$ and $k_2 = \max\{k_2^{(1)}, k_2^{(2)}\}$ then the third assertion of Lemma 9 holds for k_1 and k_2. By the definition of the points z_i, $i = 1, 2, 3$ it implies the first assertion of the lemma.

Let $\xi_n(\overline{x}_{cr}^{(1)})$ be a dynamical partition of the point $\overline{x}_{cr}^{(1)}$. According to Lemma 2 the intervals $\Delta_0^{(n)}(\overline{x}_{cr}^{(1)})$ and $\Delta_0^{(n-1)}(\overline{x}_{cr}^{(1)})$ are K-comparable. Hence, it implies that the intervals $[z_s, z_{s+1}]$, $s = 1, 2, 3$ are pairwise $K^{k_1+k_2}$-comparable. It is easy to see that the intervals $[f_1^{q_n}(z_s), f_1^{q_n}(z_{s+1})]$, $s = 1, 2, 3$ are pairwise $K^{k_1+k_2}$-comparable. Obviously,

$$\frac{1}{K^{k_0+1}} \le \frac{\ell(\Delta_0^{(n-1)}(\overline{x}_{cr}^{(1)}))}{\ell([z_1, z_2])} \le K^{k_0+1}, \qquad \frac{1}{K^{k_0+1}} \le \frac{\ell(\Delta_0^{(n-1)}(\overline{x}_{cr}^{(1)}))}{\ell([f_1^{q_n}(z_1), f_1^{q_n}(z_2)])} \le K^{k_0+1}.$$

Since the intervals $\Delta_0^{(n-1)}(\overline{x}_{cr}^{(1)})$ and $\Delta_0^{(n-1)}(f_1^{-q_{n-1}}(\overline{x}_{cr}^{(1)}))$ are K-comparable and $x_0 \in \Delta_0^{(n-1)}(f_1^{-q_{n-1}}(\overline{x}_{cr}^{(1)})) \cup \Delta_0^{(n-1)}(\overline{x}_{cr}^{(1)})$ we get $\max_{1 \le i \le 4}\{\ell([f^{q_n}(z_i), x_0]), \ell([z_i, x_0])\}$ $\le (K + 1)K^{k_0+1}\ell([z_1, z_2])$. If we take $R_1 = (K + 1)K^{k_0+k_1+k_2}$, then we obtain the proof of the second assertion of Lemma 9 with constant R_1. Lemma 9 is proved.

Proof of Lemma 10 Suppose, the triples of intervals ($[z_s, z_{s+1}]$, $s = 1, 2, 3$) and ($[h(z_s), h(z_{s+1})]$, $s = 1, 2, 3$) satisfy the conditions of Lemma 9. We want to compare the distortion $Dist(z_1, z_2, z_3, z_4; f_1^{q_n})$ and $Dist(h(z_1), h(z_2), h(z_3), h(z_4);$ $f_2^{q_n})$. We estimate only the first distortion, the second one can be estimated analogously. Obviously

$$Dist(z_1, z_2, z_3, z_4; f_1^{q_n}) = \prod_{i=0}^{q_n-1} Dist(f_1^i(z_1), f_1^i(z_2), f_1^i(z_3), f_1^i(z_4); f_1).$$

We denote

$$J_r(x_{cr}^{(1)}) = \Delta_0^{(r)}(x_{cr}^{(1)}) \cup \Delta_0^{(r-1)}(x_{cr}^{(1)}), \quad A = \{i : (f_1^i(z_1), f_1^i(z_4)) \cap J_r(x_{cr}^{(1)}) = \emptyset\},$$

$$B = \{i : (f_1^i(z_1), f_1^i(z_4)) \cap J_r(x_{cr}^{(1)}) \ne \emptyset\}.$$

It is clear that $A \cup B = \{0, 1, \ldots, q_n\}$.

Next we rewrite $Dist(z_1, z_2, z_3, z_4; f_1^{q_n})$ in the form

$$Dist(z_1, z_2, z_3, z_4; f_1^{q_n}) = \prod_{i \in A} Dist(f_1^i(z_1), f_1^i(z_2), f_1^i(z_3), f_1^i(z_4); f_1) \times$$

$$\times \prod_{i \in B} Dist(f_1^i(z_1), f_1^i(z_2), f_1^i(z_3), f_1^i(z_4); f_1). \tag{11}$$

We estimate the first factor in (11). Using the Lemmas 4 we obtain

$$|\prod_{i \in A} Dist(f_1^i(z_1), f_1^i(z_2), f_1^i(z_3), f_1^i(z_4); f_1) - 1| =$$

$$= |\prod_{i \in A}(1 + O(\ell([f_1^i(z_1), f_1^i(z_4)])^{1+\nu}) - 1| = \max_i(\ell([f_1^i(z_1), f_1^i(z_4)]))^{\nu} \times$$

$$\times O(\sum_{i \in A} \ell([f_1^i(z_1), f_1^i(z_4)])) = O(\lambda_{f_1}^{n\nu}),$$

where $\nu > 0$ and $0 < \lambda_{f_1} < 1$. We fix $\varepsilon > 0$. There exists $N_0 = N_0(\varepsilon) \geq 1$ such that for any $n \geq N_0$ the estimate

$$|\prod_{i \in A} Dist(f_1^i(z_1), f_1^i(z_2), f_1^i(z_3), f_1^i(z_4); f_1) - 1| < C_6 \varepsilon. \tag{12}$$

holds. We now estimate the second factor in (11). We rewrite the second factor in the following form

$$\prod_{i \in B} Dist(f_1^i(z_1), f_1^i(z_2), f_1^i(z_3), f_1^i(z_4); f_1) =$$

$$= \prod_{i \in B, i \neq l} Dist(f_1^i(z_1), f_1^i(z_2), f_1^i(z_3), f_1^i(z_4); f_1) \times$$

$$\times Dist(f_1^l(z_1), f_1^l(z_2), f_1^l(z_3), f_1^l(z_4); f_1). \tag{13}$$

By applying Lemmas 5 and 8 we obtain

$$|Dist(f_1^l(z_1), f_1^l(z_2), f_1^l(z_3), f_1^l(z_4); f_1) - m_1| < \frac{L}{8}. \tag{14}$$

Using Lemma 6 for the first factor in (13), we get

$$|\prod_{i \in B, i \neq l} Dist(f_1^i(z_1), f_1^i(z_2), f_1^i(z_3), f_1^i(z_4); f_1) - 1| = |\prod_{i \in B, i \neq l}(1 + O(\frac{\ell([f_1^i(z_1), f_1^i(z_4)])}{d_i})^2) - 1| =$$

$$= |\exp\{\sum_{i \in B, i \neq l} \log(1 + O(\frac{\ell([f_1^i(z_1), f_1^i(z_4)])}{d_i})^2)\} - 1| \leq const \sum_{i \in B, i \neq l} (\frac{\ell([f_1^i(z_1), f_1^i(z_4)])}{d_i})^2 =$$

$$= const \sum_{q=0}^{n-r} \sum_{i:[f_1^i(z_1), f_1^i(z_4)] \subset (J_{n-q}(x_{cr}^{(1)}) \setminus J_{n-q+1}(x_{cr}^{(1)})), i \neq l} (\frac{\ell([f_1^i(z_1), f_1^i(z_4)])}{d_i})^2$$

Obviously,

$$\sum_{i:[f_1^i(z_1), f_1^i(z_4)] \subset (J_{n-q}(x_{cr}^{(1)}) \setminus J_{n-q+1}(x_{cr}^{(1)})), i \neq l} (\frac{\ell([f_1^i(z_1), f_1^i(z_4)])}{d_i}) = const$$

and it follows from Lemma 3 that $\frac{\ell([f_1^i(z_1), f_1^i(z_4)])}{d_i} \leq const \lambda_{f_1}^{k_0+1+q}$. Consequently

$$|\prod_{i \in B, i \neq l} Dist(f_1^i(z_1), f_1^i(z_2), f_1^i(z_3), f_1^i(z_4); f_1) - 1| \leq C_7 \lambda_{f_1}^{k_0}, \quad (15)$$

where $C_7 > 0$ depends only on f_1 and $0 \leq \lambda_{f_1} \leq 1$ is defined in Lemma 3.

Similarly one can show that for the triple of intervals $([h(z_s), h(z_{s+1})]$, $s = 1, 2, 3)$ the following inequality

$$|\prod_{i \in B, i \neq l} Dist(f_2^i(h(z_1)), f_2^i(h(z_2)), f_2^i(h(z_3)), f_2^i(h(z_4)); f_2) - 1| \leq C_8 \lambda_{f_2}^{k_0}, \quad (16)$$

where $C_8 > 0$ depends only on f_2 and $0 \leq \lambda_{f_2} \leq 1$ is defined in Lemma 3.

If we choose

$$k_0 = \max\{[\log_{\lambda_{f_1}} \frac{L}{(8m_1 + L)C_7}] + 1, [\log_{\lambda_{f_2}} \frac{L}{(8m_2 + L)C_8}] + 1\},$$

then from the relations (11)–(19) it implies that for sufficiently large n

$$|Dist(z_1, z_2, z_3, z_4; f_1^{q_n}) - m_1| < \frac{L}{4}. \quad (17)$$

Similarly

$$|Dist(h(z_1), h(z_2), h(z_3), h(z_4); f_2^{q_n}) - m_2| < \frac{L}{4}. \quad (18)$$

The inequalities (17) and (18) implies

$$\frac{Dist(z_1, z_2, z_3, z_4; f_1^{q_n})}{Dist(h(z_1), h(z_2), h(z_3), h(z_4); f_2^{q_n})} - 1 \geq \frac{4(m_1 - m_2) - 2L}{4m_2 + L} > 0, \quad (19)$$

if $m_1 > m_2$, and

$$\frac{Dist(z_1, z_2, z_3, z_4; f_1^{q_n})}{Dist(h(z_1), h(z_2), h(z_3), h(z_4); f_2^{q_n})} - 1 \leq \frac{4(m_1 - m_2) + 2L}{4m_2 - L} < 0, \qquad (20)$$

if $m_1 < m_2$. If we set

$$R_2 := \min\{\frac{|4(m_1 - m_2) - 2L|}{4m_2 - L}, \frac{|4(m_1 - m_2) + 2L|}{4m_2 + L}\}, \qquad (21)$$

then it follows from (19)–(21) that the assertion of the lemma holds.

Acknowledgements The author would like to thank Professors A. A. Dzhalilov and K. M. Khanin for making several useful suggestions which improved the text of the paper.

References

1. Arnol'd, V.I.: Small denominators: I. Mappings from the circle onto itself. Izv. Akad. Nauk SSSR, Ser. Mat. **25**, 21–86 (1961)
2. Avila, A.: On rigidity of critical circle maps. Bull. Math. Soc. **44**(4), 611–619 (2013)
3. Cornfeld, I.P., Fomin, S.V., Sinai, YaG: Ergodic Theory. Springer, Berlin (1982)
4. Denjoy, A.: Sur les courbes définies par les équations différentielles à la surface du tore. J. Math. Pures Appl. **11**, 333–375 (1932)
5. Dzhalilov, A.A., Khanin, K.M.: On invariant measure for homeomorphisms of a circle with a point of break. Funct. Anal. Appl. **32**(3), 153–161 (1998)
6. de Faria, E., de Melo, W.: Rigidity of critical circle mappings. I. J. Eur. Math. Soc. (JEMS) **1**(4), 339–392 (1999)
7. Graczyk, J., Swiatek, G.: Singular measures in circle dynamics. Commun. Math. Phys. **157**, 213–230 (1993)
8. Herman, M.: Sur la conjugaison différentiable des difféomorphismes du cercle à des rotations. Inst. Hautes Etudes Sci. Publ. Math. **49**, 225–234 (1979)
9. Katznelson, Y., Ornstein, D.: The absolute continuity of the conjugation of certain diffeomorphisms of the circle. Ergod. Theor. Dyn. Syst. **9**, 681–690 (1989)
10. Khanin, K.: Universal estimates for critical circle mappings. CHAOS **2**, 181–186 (1991)
11. Khanin, K., Teplinsky, A.: Robust rigidity for circle diffeomorphisms with singularities. Invent. Math. **169**, 193–218 (2007)
12. Khanin, K.M., Sinai, YaG: Smoothness of conjugacies of diffeomorphisms of the circle with rotations. Russ. Math. Surv. **44**, 69–99 (1989). Translation of Usp. Mat. Nauk **44**, 57–82 (1989)
13. Khmelev, D., Yampolsky, M.: Rigidity problem for analytic critical circle maps. Mos. Math. J. **6**(2), 317–351 (2006)
14. Moser, J.: A rapid convergent iteration method and non-linear differential equations. II. Ann. Scuola Norm. Sup. Pisa **20**(3), 499–535 (1966)
15. Swiatek, G.: Rational rotation numbers for maps of the circle. Commun. Math. Phys. **119**(1), 109–128 (1988)
16. YoccozJ, C.: Il n'a a pas de contre-exemple de Denjoy analytique. C. R. Acad. Sci., Paris, Ser. I Math. **298**(7), 141–144 (1984)

ε-Positional Strategy in the Second Method of Differential Games of Pursuit

Tukhtasinov Muminjon

Abstract In the present work, methods of completion on any vicinity of a terminal set are offered. At the same time the type of a terminal set and a condition quite of a sweepability significantly used. The pursuing player applies ε positional strategy. At the end the work two examples illustrating the received results are given.

Keywords Differential game · Pursuer · Evader · Strategy · Methods of Pontryagin

1 Introduction

To solve effectively linear differential games of pursuit from the pursuer's perspective, several methods have been developed. In the fundamental work [1, 2] L.S. Pontryagin gave the complete description of his results on linear differential games including the second method that had a large field of applicability. The important mathematical apparatus used by L.S. Pontryagin was the apparatus of multivalued mappings. At present, this apparatus has been widely used in the theory of optimal control and in the theory of differential games. In particular, certain aspects of this method were concretized in all mathematical rigor in [3].

The main problem in the pursuit theory is selection of those points from which the pursuit can be finished in a finite time. There is a number of works, where sufficient conditions of a general type are given for possibility to finish the pursuit from the given point z^0 and the guarantee time is effectively calculated. In real conflict situations, controls are usually selected on the basis of some information about the dynamic capabilities of objects and the current changes in the system states. The players (pursuer and evader) use their awareness to achieve their goals. The pursuer goal is to remove $z(t)$ from z^0 on M for as short a time as possible, the evader has the opposite goal.

Special attention should be given paper [3], in which, at first the concept of the lower alternating integral is introduced. This integral, according to the author's pro-

T. Muminjon (✉)
Faculty of Mathematics, National University of Uzbekistan, Tashkent, Uzbekistan
e-mail: mumin51@mail.ru

© Springer Nature Switzerland AG 2018
A. Azamov et al. (eds.), *Differential Equations and Dynamical Systems*,
Springer Proceedings in Mathematics & Statistics 268,
https://doi.org/10.1007/978-3-030-01476-6_13

posal, is called dual to the alternating Pontryagin integral. Because of the lower alternating integral and the Azamov theorem [4], it was possible to solve the Pontryagin epsilon problem. Additionally, connection was established between these two alternating integrals, hereupon the question of getting into any neighborhood of the terminal set is completely resolved with the help of the epsilon-positional strategy of the pursuer [3, 4].

In works [5–7], the pursuit problem is studied for positional games: the pursuit control is constructed on the basis of the position information (t, z); in the monograph [6], in particular, it is considered the problem on reduction of generalized motions on the terminal set. In this connection, one can put the problem of allocating one motion from the mentioned generalized motions. It is easy to invent examples in which it is not satisfied the condition on a single element of the support set from [5, 7].

There are classes of strategies in the theory of differential games: quasilinear, stroboscopic, positional and others. From the point of view of the mathematical statement of the problem by indicating and constructing the strategies of the players in order to achieve the set goal, the game is considered completely solved.

In the present work, the new method is proposed consisting of two variants to complete the pursuit in any neighborhood of the terminal set under additional conditions on the game parameters. If the terminal set is significantly used in the first variant, then the sweepability condition is used in the second one. Moreover, the pursuer uses ε-positional strategy to complete the pursuit [3]. At the end of the work, two examples are given that illustrating the obtained results.

2 Preliminaries

It is considered the linear differential game described by the equation

$$\dot{z} = Cz - u + v \tag{1}$$

where $z \in \mathbb{R}^d$ is a phase vector; C is a constant square matrix of the order $d \times d$; $u \in P, v \in Q$ are control parameters, moreover P, Q are nonempty compact subsets of the space $\mathbb{R}^d$. The terminal set on which the game is finished is a nonempty closed subset M of the space $\mathbb{R}^d$.

Equation (1), sets P, Q, and M describe the two-player differential game: the pursuer that controls the vector u and the evader that controls the vector v. The motion of the point z begins at $t = 0$ and proceeds under the action of measurable controls $u(t) \in P$ and $v(t) \in Q, t > 0$.

As it is known, the alternating Pontryagin integral [1]

$$W_2(\tau) = \int\limits_{M,0}^{\tau} \left[e^{sC} P \, ds \overset{*}{-} e^{sC} Q \, ds \right]$$

is the limit of the alternating sum of sequences of compact convex sets

$$U_1, U_2, \ldots, U_n \quad \text{and} \quad V_1, V_2, \ldots, U_n$$

with the initial value $A_0 = M$; $U_i = \int_{\tau_{i-1}}^{\tau_i} e^{sC} P \, ds$, $V_i = \int_{\tau_{i-1}}^{\tau_i} e^{sC} Q \, ds$, $i = 1, \ldots, n$;

where $0 = \tau_0 < \tau_1 < \cdots < \tau_n = \tau$ is a partition of $[0, \tau]$. The sum is determined as follows [1]:

$$A_i = (A_{i-1} + U_i) \overset{*}{-} V_i, \quad i = 1, \ldots, n.$$

A_n will be written in the expanded form by the formula

$$A_n = (\ldots ((((M + U_1) \overset{*}{-} V_1) + U_2) \overset{*}{-} V_2) + \ldots + U_n) \overset{*}{-} V_n. \tag{2}$$

We say that the end of the pursuit game from the starting state $z(0) = z^0$ is possible if the pursuer has such admissible strategy $U(z^0, t, v)$ that for any admissible control of the evader $v(t) \in Q$ and $u[t] = U(z^0, t, v(t)) \in P$, $t \geq 0$, the end of the game takes place on M, i.e. there exists a finite time moment $\tau > 0$ such that $z(\tau) \in M$ where $z(t)$, $t \geq 0$, is the corresponding solution of the problem (1) at $u = u[t]$, $v = v(t)$, $t \geq 0$ and $z(0) = z^0$.

Theorem 1 [1] *If the following inclusion*

$$e^{\tau C} z^0 \in W_2(\tau)$$

holds for a given starting point $z^0 \notin M$ at a time moment τ, then the pursuit from the point z^0 can be completed in the time τ.

Let M be a closed convex set, F and G be compact convex sets. Then the following easily proved inclusions are true [1]:

$$(M \overset{*}{-} F) + G \subset (M + G) \overset{*}{-} F. \tag{3}$$

$$(M \overset{*}{-} F) \overset{*}{-} G = M \overset{*}{-} (F + G). \tag{4}$$

If we put

$$U = U_1 + \cdots + U_n, \quad V = V_1 + \cdots + V_n,$$

then using formulas (3), (4), by virtue of additivity of the integral we obtain from (2):

$$A_n \subset \left(M + \int_0^\tau e^{sC} P \, ds \right) \overset{*}{-} \int_0^\tau e^{sC} Q \, ds,$$

what implies

$$W_2(\tau) \subset \left(M + \int\limits_0^\tau e^{sC} P \, ds \right) \overset{*}{-} \int\limits_0^\tau e^{sC} P \, ds = \overline{W}(\tau).$$

3   Main results

Let $\varepsilon > 0$, $W \subset \mathbb{R}^d$. We denote by $L(\varepsilon, W)$ the totality of all measurable functions $\omega(\cdot) : [0, \varepsilon] \to W$ [3].

Definition 1 The mapping

$$P_\varepsilon : \mathbb{R}^d \times \mathbb{R}^d \times \mathbb{R}^d \to L(\varepsilon, P)$$

is said to be the ε-positional strategy of the pursuer.

Definition 2 The mapping

$$Q_\varepsilon : \mathbb{R}^d \to L(\varepsilon, Q)$$

is said to be the ε-positional strategy of the evader.

Definition 3 We say that the game (1) from the starting point z^0 is finished by the time moment τ with the ε-positional strategy if there exists an ε-positional strategy of the pursuer $P_\varepsilon(\cdot, \cdot, \cdot)$ such that for any ε-positional strategy $Q_\varepsilon(\cdot)$ of the evader the inclusion $z(\tau) \in M$ takes place.

Assumption 1 For arbitrary numbers $0 < a \le b$, $t \ge 0$, the inclusion

$$\int\limits_a^b e^{-sC} P \, ds \subset \int\limits_a^b e^{-(t+s)C} P \, ds$$

holds.

Assumption 2 Let for any $\varepsilon = \tau/k$, there be a partition $M_1, M_2, \ldots, M_k$ of the set M such that

$$\overline{W}(\tau) \subset \sum_{i=1}^k \left(\left(M_{k-i+1} + \int\limits_{(i-1)\varepsilon}^{i\varepsilon} e^{sC} P \, ds \right) \overset{*}{-} \int\limits_{(i-1)\varepsilon}^{i\varepsilon} e^{sC} Q \, ds \right).$$

Suppose the following means hypothetical inclusion

$$e^{\tau C} z^0 \in \overline{W}(\tau). \tag{5}$$

Theorem 2 *Let for a given starting state z^0 and a time moment τ, the inclusion (5) and Assumptions 1 and 2 be valid. Then for any number $\alpha > 0$, there is a positive number ε such that the game (1) will be finished by the time moment τ with the ε-positional strategy in the α-neighborhood of the set $M : z(\tau) \in M_\alpha$.*

Proof Let for a given starting state z^0 and a time moment τ the inclusion (5) hold. Evidently, for any positive number α, there exists a natural number k such that the inequality

$$\left| \int_0^\varepsilon e^{sC} P \, ds \right| < \frac{\alpha}{2} \tag{6}$$

holds for $\varepsilon = \tau/k$ where $|F| = \max\{\|f\| : f \in F\}$ for a compact set F.
    Since

$$\int_{(i-1)\varepsilon}^{i\varepsilon} e^{sC} P \, ds = \int_{\tau-i\varepsilon}^{\tau-(i-1)\varepsilon} e^{(\tau-s)C} P \, ds \quad \text{and} \quad \int_{(i-1)\varepsilon}^{i\varepsilon} e^{sC} Q \, ds = \int_{\tau-i\varepsilon}^{\tau-(i-1)\varepsilon} e^{(\tau-s)C} Q \, ds$$

then using Assumption 2, we obtain from (5)

$$e^{\tau C} z^0 \in \sum_{i=1}^k \left(\left(M_i + \int_{(i-1)\varepsilon}^{i\varepsilon} e^{(\tau-s)C} P \, ds \right) \overset{*}{-} \int_{(i-1)\varepsilon}^{i\varepsilon} e^{(\tau-s)C} Q \, ds \right). \tag{7}$$

    Thus, by virtue of the inclusion (7), for the given starting state z^0 and the time moment τ there exist vectors $g^i \in \left(e^{-\tau C} M_i + \int_{(i-1)\varepsilon}^{i\varepsilon} e^{-sC} P \, ds \right) \overset{*}{-} \int_{(i-1)\varepsilon}^{i\varepsilon} e^{-sC} Q \, ds$, $i = 1, 2, \ldots, k$, for which the equality

$$z^0 = g^1 + g^2 + \cdots + g^k \tag{8}$$

is valid.
    Now let $v(s) \in Q, 0 \leq s \leq \tau$ be arbitrary admissible control of the evader. On the segment $[0, \varepsilon]$, the pursuer chooses arbitrary admissible control $u^{(1)}(s), 0 \leq s \leq \varepsilon$. Then at the time moment ε we have the position of the form

$$z(\varepsilon) = e^{\varepsilon C} z^0 - \int_0^\varepsilon e^{(\varepsilon-s)C} \left(u^{(1)}(s) - v(s) \right) ds,$$

what implies the equality

$$z(\varepsilon) - e^{\varepsilon C}\left(z^0 - g^1 - \int_0^\varepsilon e^{-sC}u^{(1)}(s)\,ds\right) = e^{\varepsilon C}\left(g^1 + \int_0^\varepsilon e^{-sC}v(s)\,ds\right). \quad (9)$$

By Assumption 1, we have

$$g^1 + \int_0^\varepsilon e^{-sC}v(s)\,ds \in g^1 + \int_0^\varepsilon e^{-sC}Q\,ds \subset e^{-\tau C}M_1 + \int_0^\varepsilon e^{-sC}P\,ds \subset e^{-\tau C}M_1$$

$$+ \int_\varepsilon^{2\varepsilon} e^{-sC}P\,ds.$$

So, we obtain from (9)

$$z(\varepsilon) - e^{\varepsilon C}\left(z^0 - g^1 - \int_0^\varepsilon e^{-sC}u^{(1)}(s)\,ds\right) \in e^{\varepsilon C}\left(e^{-\tau C}M_1 + \int_\varepsilon^{2\varepsilon} e^{-sC}P\,ds\right).$$

Hence, knowing $z(\varepsilon)$ and $z^1 = z^0 - g^1$, one can find an element $m_1 \in M_1$ and admissible control $u^{(2)}(s) \in P$, $\varepsilon \leq s \leq 2\varepsilon$ such that the equalities

$$z(\varepsilon) - e^{\varepsilon C}\left(z^0 - g^1 - \int_0^\varepsilon e^{-sC}u^{(1)}(s)\,ds\right) = e^{\varepsilon C}\left(e^{-\tau C}m_1 + \int_\varepsilon^{2\varepsilon} e^{-sC}u^{(2)}(s)\,ds\right)$$

and

$$g^1 + \int_0^\varepsilon e^{-sC}v(s)\,ds = e^{-\tau C}m_1 + \int_\varepsilon^{2\varepsilon} e^{-sC}u^{(2)}(s)\,ds$$

hold.

Moreover, the trajectory $z(t)$ on the segment $[0, \varepsilon]$ is determined as the solution of the Cauchy problem

$$\dot{z} = Cz - u^{(1)}(t) + v(t), \quad z(0) = z^0$$

where $v(\cdot) = Q_\varepsilon(z(0))$, $u^{(1)}(\cdot) = P_\varepsilon(z(0), z^0, 0)$. In addition, $u^{(1)}(\cdot)$ is arbitrary but fixed element of the space $L(\varepsilon, P)$.

Then we apply the method of mathematical induction. Let $1 < i < k$. Using the explicit form of the solution of the system (1), write its value by the moment $i\varepsilon$:

$$z(i\varepsilon) = e^{i\varepsilon C} z^0 - \int\limits_0^{i\varepsilon} e^{(i\varepsilon - s)C}\Big(u(s) - v(s)\Big)\, ds.$$

We obtain after not difficult transformations

$$z(i\varepsilon) - e^{i\varepsilon C}\left(z^0 - g^1 - g^2 - \cdots - g^i + g^1 + \int\limits_0^\varepsilon e^{-sC} v(s)\, ds + g^2 + \int\limits_\varepsilon^{2\varepsilon} e^{-sC} v(s)\, ds + \cdots + \right.$$

$$g^{i-1} + \int\limits_{(i-2)\varepsilon}^{(i-1)\varepsilon} e^{-sC} v(s)\, ds - \int\limits_0^\varepsilon e^{-sC} u^{(1)}(s)\, ds - \int\limits_\varepsilon^{2\varepsilon} e^{-sC} u^{(2)}(s)\, ds - \int\limits_{2\varepsilon}^{3\varepsilon} e^{-sC} u^{(3)}(s)\, ds - \cdots -$$

$$\left. \int\limits_{(i-1)\varepsilon}^{i\varepsilon} e^{-sC} u^{(i)}(s)\, ds \right) = e^{i\varepsilon C}\left(g^i + \int\limits_{(i-1)\varepsilon}^{i\varepsilon} e^{-sC} v(s)\, ds \right). \tag{10}$$

Since

$$g^i + \int\limits_{(i-1)\varepsilon}^{i\varepsilon} e^{-sC} v(s)\, ds \in g^i + \int\limits_{(i-1)\varepsilon}^{i\varepsilon} e^{-sC} Q\, ds \subset e^{-\tau C'} M_i + \int\limits_{(i-1)\varepsilon}^{i\varepsilon} e^{-sC} P\, ds \tag{11}$$

then using Assumption 1, we obtain the inclusion

$$e^{-\tau C} M_i + \int\limits_{(i-1)\varepsilon}^{i\varepsilon} e^{-sC} v(s)\, ds \in e^{-\tau C} M_i + \int\limits_{i\varepsilon}^{(i+1)\varepsilon} e^{-sC} P\, ds. \tag{12}$$

Thus, (10)–(12) imply that there exists an element $m_i \in M_i$ and admissible control $u^{(i+1)}(s) \in P$, $i\varepsilon \le s \le (i+1)\varepsilon$ such that:

$$z(i\varepsilon) - e^{i\varepsilon C}\left(z^0 - g^1 - g^2 - \cdots - g^i + e^{-\tau C}(m_1 + \cdots + m_{i-1}) - \int\limits_0^\varepsilon e^{-sC} u^{(1)}(s)\, ds \right) =$$

$$e^{i\varepsilon C}\left(e^{-\tau C} m_i + \int\limits_{i\varepsilon}^{(i+1)\varepsilon} e^{-sC} u^{(i+1)}(s)\, ds \right)$$

and

$$g^i + \int\limits_{(i-1)\varepsilon}^{i\varepsilon} e^{-sC} v(s)\, ds = e^{-\tau C} m_i + \int\limits_{i\varepsilon}^{(i+1)\varepsilon} e^{-sC} u^{(i+1)}(s)\, ds.$$

Introduce the following notations: $z^i = z^0 - g^1 - \cdots - g^i + e^{-\tau C}(m_1 + \cdots + m_{i-1})$, $m_0 = 0$, and put: $P_\varepsilon(z(i\varepsilon), z^i, m_i)(t - i\varepsilon) = u^{(i+1)}(t)$, $i\varepsilon \leq t < (i+1)\varepsilon$.

If we repeat all induction arguments for the case of $i = k$, then we obtain from (10) and (11)

$$z(k\varepsilon) - e^{k\varepsilon C}\left(z^0 - g^1 - g^2 - \cdots - g^k + e^{-\tau C}(m_1 + \cdots + m_{k-1}) - \int_0^\varepsilon e^{-sC} u^{(1)}(s)\, ds\right) =$$

$$e^{k\varepsilon C}\left(g^k + \int_{(k-1)\varepsilon}^{k\varepsilon} e^{-sC} v(s)\, ds\right) \in e^{k\varepsilon C}\left(g^k + \int_{(k-1)\varepsilon}^{k\varepsilon} e^{-sC} Q\, ds\right) \subset$$

$$e^{k\varepsilon C}\left(e^{-\tau C} M_k + \int_{(k-1)\varepsilon}^{k\varepsilon} e^{-sC} P\, ds\right).$$

The last inclusion with the equality (8) imply

$$z(\tau) \in M - \int_0^\varepsilon e^{(\tau-s)C} P\, ds + \int_{(k-1)\varepsilon}^\tau e^{(\tau-s)C} P\, ds.$$

Since the following easily proved inclusions

$$\int_0^\varepsilon e^{(\tau-s)C} P\, ds \subset \int_0^\varepsilon e^{(\varepsilon-s)C} P\, ds = \int_0^\varepsilon e^{sC} P\, ds \quad \text{and} \quad \int_{(k-1)\varepsilon}^\tau e^{(\tau-s)C} P\, ds = \int_0^\varepsilon e^{sC} P\, ds,$$

take place, then, taking into account the inequality (6), we obtain

$$z(\tau) \in M_\alpha.$$

Let's consider a class of the theory of linear differential pursuit games.

Definition 4 Let F and G be arbitrary subsets of the space $\mathbb{R}^d$. We say that the set G sweeps out the set F if the equality $(F \stackrel{*}{-} G) + G = F$ holds.

Lemma 1 ([8]) *Let F, G, K be nonempty convex closed sets from $\mathbb{R}^d$, moreover G, K be bounded sets and $(F \stackrel{*}{-} G) + G = F$, i.e. the set G sweeps out the set F. Then*

$$(K + F) \stackrel{*}{-} G = K + (F \stackrel{*}{-} G)$$

and the set G sweeps out the set $F + K$.

Assumption 3 For any $t \in [0, \tau]$, the set $\int\limits_0^t e^{-sC} Q$ sweeps out the set $\int\limits_0^t e^{-sC} P$.

Theorem 3 *Let for a given starting state z^0 and a time moment τ the inclusion (5) and Assumptions 1 and 3 hold. Then for any number $\alpha > 0$ there exists a positive number ε such that the game (1) is finished with the ε-positional strategy by the time moment τ in the α-neighborhood of the set $M : z(\tau) \in M_\alpha$.*

Proof Choose a positive number $\varepsilon > 0$. By Assumption 3, there exists a set $D \subset \mathbb{R}^d$ such that

$$D + \int\limits_0^\varepsilon e^{-sC} Q \, ds = \int\limits_0^\varepsilon e^{-sC} P \, ds,$$

what implies

$$e^{-\varepsilon C} D + \int\limits_\varepsilon^{2\varepsilon} e^{-sC} Q \, ds = \int\limits_\varepsilon^{2\varepsilon} e^{-sC} P \, ds.$$

Once again applying to both parts of this equality the operator $e^{-\varepsilon C}$, we obtain

$$e^{-2\varepsilon C} D + \int\limits_{2\varepsilon}^{3\varepsilon} e^{-sC} Q \, ds = \int\limits_{2\varepsilon}^{3\varepsilon} e^{-sC} P \, ds,$$

and so on. Hence, for any $i \geq 1$, we have

$$e^{-i\varepsilon C} D + \int\limits_{i\varepsilon}^{(i+1)\varepsilon} e^{-sC} Q \, ds = \int\limits_{i\varepsilon}^{(i+1)\varepsilon} e^{-sC} P \, ds. \tag{13}$$

If we set $\varepsilon = \tau/k$ for any natural number k, then summing (13) by i from 0 to $k-1$, we obtain on one side

$$(E + e^{-\varepsilon C} + \cdots + e^{-(k-1)\varepsilon C}) D + \int\limits_0^{k\varepsilon} e^{-sC} Q \, ds = \int\limits_0^{k\varepsilon} e^{-sC} P \, ds, \tag{14}$$

on the other hand, referring to Assumption 1, we obtain from (13)

$$(E + e^{-\varepsilon C} + \cdots + e^{-(k-1)\varepsilon C}) D = \sum_{i=1}^{k} \left(\int\limits_{(i-1)\varepsilon}^{i\varepsilon} e^{-sC} P \, ds \overset{*}{-} \int\limits_{(i-1)\varepsilon}^{i\varepsilon} e^{-sC} Q \, ds \right). \tag{15}$$

Comparing equalities (14) and ((15), we get

$$\int_0^{k\varepsilon} e^{-sC} P\, ds \overset{*}{-} \int_0^{k\varepsilon} e^{-sC} Q\, ds = \sum_{i=1}^{k}\left(\int_{(i-1)\varepsilon}^{i\varepsilon} e^{-sC} P\, ds \overset{*}{-} \int_{(i-1)\varepsilon}^{i\varepsilon} e^{-sC} Q\, ds \right)$$

or

$$\int_0^{k\varepsilon} e^{(\tau-s)C} P\, ds \overset{*}{-} \int_0^{k\varepsilon} e^{(\tau-s)C} Q\, ds = \sum_{i=1}^{k}\left(\int_{(i-1)\varepsilon}^{i\varepsilon} e^{(\tau-s)C} P\, ds \overset{*}{-} \int_{(i-1)\varepsilon}^{i\varepsilon} e^{(\tau-s)C} Q\, ds \right).$$

Using the statement of Lemma and Assumption 3 on complete sweepability, we obtain from here:

$$\left(M + \int_0^{k\varepsilon} e^{(\tau-s)C} P\, ds \right) \overset{*}{-} \int_0^{k\varepsilon} e^{(\tau-s)C} Q\, ds =$$

$$= M + \sum_{i=1}^{k}\left(\int_{(i-1)\varepsilon}^{i\varepsilon} e^{(\tau-s)C} P\, ds \overset{*}{-} \int_{(i-1)\varepsilon}^{i\varepsilon} e^{(\tau-s)C} Q\, ds \right). \tag{16}$$

Suppose that for a time moment τ the inclusion (5) and Assumptions 1 and 3 hold. It is clear that if $u(s) \in P$, $v(s) \in Q$, $0 \le s \le t$ are arbitrary admissible controls of the pursuer and evader, respectively, then after substituting them into the right-hand side of (1), we obtain the inhomogeneous system of linear differential equations

$$\dot{z} = Cz - u(t) + v(t),$$

the solution of which with the initial condition $z(0) = z^0$ is presented by the Cauchy formula:

$$z(t) = e^{tC} z^0 - \int_0^t e^{(t-s)C}\big(u(s) - v(s)\big)\, ds. \tag{17}$$

For a given positive number α, choose a natural number k such that the inequality (6) takes place.

It follows from (5) and (16) that for a given starting state z^0 and τ, there exist elements $m \in M$ and $g^i \in \int_{(i-1)\varepsilon}^{i\varepsilon} e^{-sC} P\, ds \overset{*}{-} \int_{(i-1)\varepsilon}^{i\varepsilon} e^{-sC} Q\, ds$, $i = 1, 2, \ldots, k$, for which the equality

$$z^0 = e^{-\tau C} m + g^1 + g^2 + \cdots + g^k \tag{18}$$

hold.

Now let $Q_\varepsilon(\cdot)$ be arbitrary ε-positional strategy of the evader. The pursuer chooses on the segment $[0, \varepsilon]$ arbitrary admissible control $u^{(1)}(s)$, $0 \le s \le \varepsilon$. Then we have at the time moment ε the position of the form

$$z(\varepsilon) = e^{\varepsilon C} z^0 - \int_0^\varepsilon \left(e^{(\varepsilon-s)C} u^{(1)}(s) - e^{(\varepsilon-s)C} v(s) \right) ds.$$

We get from here the following equality

$$z(\varepsilon) - e^{\varepsilon C}\left(z^0 - g^1 - \int_0^\varepsilon e^{-sC} u^{(1)}(s)\, ds \right) = e^{\varepsilon C}\left(g^1 + \int_0^\varepsilon e^{-sC} v(s)\, ds \right). \quad (19)$$

By virtue of Assumption 1 we have

$$g^1 + \int_0^\varepsilon e^{-sC} v(s)\, ds \in g^1 + \int_0^\varepsilon e^{-sC} Q\, ds \subset e^{-\varepsilon C} \int_0^\varepsilon e^{-sC} P\, ds \subset \int_\varepsilon^{2\varepsilon} e^{-sC} P\, ds.$$

Thus, we obtain from (19)

$$z(\varepsilon) - e^{\varepsilon C}\left(z^0 - g^1 - \int_0^\varepsilon e^{-sC} u^{(1)}(s)\, ds \right) = e^{\varepsilon C}\left(g^1 + \int_0^\varepsilon e^{-sC} v(s)\, ds \right) \in e^{\varepsilon C} \int_\varepsilon^{2\varepsilon} e^{-sC} P\, ds.$$

Hence, knowing $z(\varepsilon)$ and $z^0 - g^1$, one can construct the admissible control $u^{(2)}(s) \in P$, $0 \le s \le \varepsilon$ such that the equalities

$$z(\varepsilon) - e^{\varepsilon C}\left(z^0 - g^1 - \int_0^\varepsilon e^{-sC} u^{(1)}(s)\, ds \right) = \int_\varepsilon^{2\varepsilon} e^{(\varepsilon-s)C} u^{(2)}(s)\, ds.$$

and

$$g^1 + \int_0^\varepsilon e^{-sC} v(s)\, ds = \int_\varepsilon^{2\varepsilon} e^{-sC} u^{(2)}(s)\, ds$$

hold where $v(\cdot) = Q_\varepsilon(z(0))$, $u^{(2)} = P_\varepsilon(z(0), z^0, 0)$.

Further we apply the method of mathematical induction. Let $1 < i < k$. Using the explicit form (17) of the solution of (1), write its value by the moment $i\varepsilon$:

$$z(i\varepsilon) = e^{i\varepsilon C} z^0 - \int_0^{i\varepsilon} \left(e^{(i\varepsilon-s)C} u(s) - e^{(i\varepsilon-s)C} v(s) \right) ds.$$

We get after not difficult transformations

$$z(i\varepsilon) - e^{i\varepsilon C}\left(z^0 - g^1 - g^2 - \cdots - g^i + g^1 + \int_0^\varepsilon e^{-sC}v(s)\,ds + g^2 + \int_\varepsilon^{2\varepsilon} e^{-sC}v(s)\,ds + \cdots +\right.$$

$$g^{i-1} + \int_{(i-2)\varepsilon}^{(i-1)\varepsilon} e^{-sC}v(s)\,ds - \int_0^\varepsilon e^{-sC}u^{(1)}(s)\,ds - \int_\varepsilon^{2\varepsilon} e^{-sC}u^{(2)}(s)\,ds + \int_{2\varepsilon}^{3\varepsilon} e^{-sC}u^{(3)}(s)\,ds + \cdots +$$

$$\left.\int_{(i-1)\varepsilon}^{i\varepsilon} e^{-sC}u^{(i)}(s)\,ds\right) = e^{i\varepsilon C}\left(g^i + \int_{(i-1)\varepsilon}^{i\varepsilon} e^{-sC}v(s)\,ds\right), \tag{20}$$

what implies by virtue of Assumption 1

$$g^i + \int_{(i-1)\varepsilon}^{i\varepsilon} e^{-sC}v^{(0)}(s)\,ds \subset g^i + \int_{(i-1)\varepsilon}^{i\varepsilon} e^{-sC}Q\,ds \subset e^{-\varepsilon C}\int_{(i-1)\varepsilon}^{i\varepsilon} e^{-sC}P\,ds \subset \int_{i\varepsilon}^{(i+1)\varepsilon} e^{-sC}P\,ds. \tag{21}$$

So, it follows from (20) and (21) that there exists an admissible control $u^{(i+1)}(s) \in P$, $i\varepsilon \le s \le (i+1)\varepsilon$ such that:

$$z(i\varepsilon) - e^{i\varepsilon C}\left(z^0 - g^1 - g^2 - \cdots - g^i - \int_0^\varepsilon e^{-sC}u^{(1)}(s)\,ds\right) = e^{i\varepsilon C}\int_{i\varepsilon}^{(i+1)\varepsilon} e^{-sC}u^{(i+1)}(s)\,ds$$

and

$$g^i + \int_{(i-1)\varepsilon}^{i\varepsilon} e^{-sC}v(s)\,ds = \int_{i\varepsilon}^{(i+1)\varepsilon} e^{-sC}u^{(i+1)}(s)\,ds$$

where $v(s) = Q_\varepsilon(z(i\varepsilon))(s - i\varepsilon)$ and $u^{(i+1)} = P_\varepsilon(z(i\varepsilon), z^i)(s - i\varepsilon)$, $i\varepsilon \le (i+1)\varepsilon$.

If we repeat all induction arguments for the case of $i = k$, then we obtain from (20) and (21)

$$z(k\varepsilon) - e^{k\varepsilon C}\left(z^0 - g^1 - g^2 - \cdots - g^k - \int_0^\varepsilon e^{-sC}u^{(1)}(s)\,ds\right) = e^{k\varepsilon C}\left(g^k + \int_{(k-1)\varepsilon}^{k\varepsilon} e^{-sC}v(s)\,ds\right) \in$$

$$e^{k\varepsilon C}\left(g^k + \int_{(k-1)\varepsilon}^{k\varepsilon} e^{-sC}Q\,ds\right) \subset e^{k\varepsilon C}\int_{(k-1)\varepsilon}^{k\varepsilon} e^{-sC}P\,ds.$$

We get from the last inclusion and the equality

$$z(k\varepsilon) - m \in -\int_0^\varepsilon e^{(k\varepsilon-s)C} P \, ds + \int_{(k-1)\varepsilon}^{k\varepsilon} e^{(k\varepsilon-s)C} P \, ds.$$

Thus,

$$\|z(k\varepsilon) - m\| \leq \left| \int_0^\varepsilon e^{(k\varepsilon-s)C} P \, ds \right| + \left| \int_{(k-1)\varepsilon}^{k\varepsilon} e^{(k\varepsilon-s)C} P \, ds \right|.$$

Now, taking into account the inequality (6), we obtain

$$\|z(\tau) - m\| \leq \left| \int_0^\varepsilon e^{(\tau-s)C} P \, ds \right| + \left| \int_{(k-1)\varepsilon}^{k\varepsilon} e^{(\tau-s)C} P \, ds \right| < \frac{\alpha}{2} + \frac{\alpha}{2} = \alpha.$$

4 Examples

1. Let $C = \lambda E, \lambda \leq 0, P = \rho K, Q = \sigma K, M = K, \rho > \sigma$ where K is the unit cube with the center at zero of the space $\mathbb{R}^d$; E be the unit matrix of the order d. Then all conditions of Theorem 1 are fulfilled, so the pursuit can be finished from any starting state $z^0 \in \mathbb{R}^d$.

2. Let in Example (1.1), $C = \lambda E, \ \lambda \leq 0, \ P = S_\rho(0), \ Q = S_\sigma(0), \ M = S_1(0),$ $\rho > \sigma$ where $S_r(0)$ is a ball of a radius $r > 0$ with the center at zero of the space $\mathbb{R}^d$; E is the square unit matrix of the order d. Then one can easily check that conditions of Assumptions 1 and 3 are fulfilled for any $z^0 \in \mathbb{R}^d$.

3. *The Pontryagin taste example.* Let

$$C = \begin{pmatrix} 0 & 1 & -1 \\ 0 & -\alpha & 0 \\ 0 & 0 & -\beta \end{pmatrix}$$

in the game (1) where α, β are positive numbers; $u = (0, -\bar{u}, 0)^T, \ v = (0, 0, \bar{v})^T,$ under the condition $\|\bar{u}\| \leq \rho, \|\bar{v}\| \leq \sigma$. The terminal set is a closed convex cylinder in the space $\mathbb{R}^{3n}$. One can show, if inequalities $\rho > \sigma, \ \rho/\alpha > \sigma/\beta$ hold, then all conditions of Assumptions 1 and 3 are valid for any $z^0 \in \mathbb{R}^d$.

References

1. Pontryagin, L.S.: Linear differential games of pursuit. Mat. Sb. **112**(154):3(7), 307–330 (1980)
2. Pontryagin, L.S.: Selected scientific works, 2, Nauka, M., 576 p (1988). (Russian)

3. Azamov, A.: On Pontryagin's second method in linear differential games of pursuit. Math. USSR-Sb. **46**(3), 429–437 (1983)
4. Azamov, A.: Semistability and duality in the theory of pontryagin alternating intefral. Sov. Math. Dokl. **37**(2), 355–359 (1988)
5. Satimov, N.: On the pursuit problem for position in differential games. Sov. Math. Dokl. **229**(4), 808–811 (1976)
6. Krasovskiy, N.N., Subbotin, A.I.: Positional differential games, Nauka, M., 455 p (1974) (Russian)
7. Pshenichny, B.N,: Linear differential games. Avtomat. i Telemekh., no. 1, pp. 65–78 (1968)
8. Satimov, N.: On the pursuit problem in linear differential games. Differ. Uravn. **9**(11), 2000–2009 (1973)

Unilateral Ball Potentials on Generalized Lebesgue Spaces with Variable Exponent

Yakhshiboev Makhmadiyor

Abstract The theorem on the boundedness of the unilateral ball potentials operator in the Lebesgue spaces with variable exponent.

Keywords Unilateral ball potentials · Riesz potential operator · Generalized Lebesgue space · Maximal function

1 Introduction

Lebesgue spaces with variable exponents have been of interest during the last years (see, for instance, [1, 3, 4, 8, 10–17, 24, 28] for the basic properties). In particular, there was an important progress concerning to the study of classical operators of Harmonic Analysis in these spaces. We refer to [9, 12, 22–24] for details on the development of this theory and [4], where the boundedness of various operators was obtained by extrapolation techniques.

Investigations in this area are strongly stimulated by applications in various problems related to objects with non-standard variable local growth (in elasticity theory, fluid mechanics, differential equations, see for example [6, 21, 27, 29, 30]). The spaces $L^{p(\cdot)}(\Omega)$ are proved to be an appropriate tool applicable in this area.

In papers [19, 25] were considered unilateral ball potentials $B_{\pm}^{\alpha}\varphi$—multidimensional analogs of operators of Riemann–Liouville fractional integration. Number of properties of unilateral ball potentials are describe in [19, 25], (see, also, [26, 29]).

In this paper, we study unilateral ball potentials of variable order in variable exponent Lebesgue spaces. We will show the boundedness of unilateral ball potentials in variable exponent Lebesgue spaces. Among the challenging problems is the Sobolev type theorem on boundedness of the unilateral ball potentials operators $B_{\pm}^{\alpha}\varphi$, from $L^{p(\cdot)}(\Omega)$ into $L^{q(\cdot)}(\Omega)$.

This paper is structured as follows. Notations and basic definitions on variable exponent spaces on metric measure spaces are given in Sect. 1. In Sect. 2 fractional

Y. Makhmadiyor (✉)
National University of Uzbekistan, Tashkent, Uzbekistan
e-mail: yahshiboev@rambler.ru

© Springer Nature Switzerland AG 2018
A. Azamov et al. (eds.), *Differential Equations and Dynamical Systems*,
Springer Proceedings in Mathematics & Statistics 268,
https://doi.org/10.1007/978-3-030-01476-6_14

maximal functions are described. The Riesz potential operator on variable Lebesgue spaces are studied in Sect. 3. In Sect. 3 there are presented some known results connected to the unilateral ball potentials. The main results, the Sobolev-type Theorems 3 and 5, are proved in Sect. 4.

Notation. Ω is a measurable set in R^n, $|\Omega|$ is its Lebesgue measure; $B(x, \delta)$ is the ball centered at x and of radius δ; $\chi_\Omega(x)$ is the characteristic function of a set Ω in R^n; $c_{n,\alpha} = \frac{\Gamma(\frac{n-\alpha}{2})2^{-\alpha}}{\pi^{n/2}\Gamma(\frac{\alpha}{2})}$; $\gamma_{n,\alpha} = \frac{\Gamma(\frac{n}{2})}{\pi^{n/2}\Gamma(\alpha)}$, $P(x) = \begin{cases} (p')_-, & |x| \geq 1, \\ (p')_+, & |x| \leq 1. \end{cases}$

2 Preliminaries

2.1 $L^{p(\cdot)}$ Spaces with Variable Exponents

We refer to [1, 5–9, 16] for details on variable Lebesgue spaces over domains in R^n, but give some necessary definitions. For a measurable function $p : \Omega \to [1, \infty)$, where $\Omega \subset R^n$ is an open set, we define $p^- := ess \inf_{x \in \Omega} p(x)$, $p^+ := ess \sup_{x \in \Omega} p(x)$.

Definition 1 By $P(\Omega)$ we denote the set of functions $p : \Omega \to [1, \infty)$ satisfying the conditions $1 < p^- \leq p(x) \leq p^+ < \infty$ on Ω

$$p(x) - p(y)| \leq \frac{A}{ln\frac{1}{|x-y|}} \quad \text{for some} \quad (x, y) \in \Omega, \quad \text{with} \quad |x - y| \leq \frac{1}{2}, \quad (1)$$

where $C > 0$ dose not depend on x and y.

In the case when $\Omega = R^n$ we have the following definition

Definition 2 Let Ω be an unbounded set. By $P_\infty(\Omega)$ we denote the set of all bounded measurable functions $p : \Omega \to [1, \infty)$, such that $1 < p^- \leq p(x) \leq p^+ < \infty$ and

$$|p(x) - p(\infty)| \leq \frac{C}{ln(e + |x|)} \quad \text{for some} \quad (x, y) \in \Omega, p(\infty) = \lim_{x \to \infty} p(x), \quad (2)$$

where $C > 0$ does not depend on x and y.

Definition 3 We define the following class of variable exponents

$$P^{log}(\Omega) := \left\{ p \in P(\Omega) : \frac{1}{p} \text{ is globally log-Holder continuous} \right\}.$$

By $c_{log}(p)$ or c_{log} we denote the log-Holder constant of $\frac{1}{p}$.

The generalized Lebesgue space $L^{p(\cdot)}(\Omega)$ with variable exponent is defined as the set of functions f on Ω for which

$$A_p(f) := \int\limits_{\Omega} |f(x)|^{p(x)} dx < \infty,$$

and equipped with the norm

$$\|f\|_{L^{p(\cdot)}(\Omega)} = \inf\left\{\lambda > 0 : A_p(\frac{f(x)}{\lambda}) \leq 1\right\}.$$

$L^{p(\cdot)}(\Omega)$ is a Banach space when $p^+ < \infty$.

If $p \in P(\Omega)$, then we define $p' \in P(\Omega)$ by $\frac{1}{p(x)} + \frac{1}{p'(x)} = 1$.

The function p' is called the dual variable exponent of p.

The notation $L^{p(\cdot)}(\Omega, \rho)$ will stand for the corresponding weighted space

$$L^{p(\cdot)}(\Omega, \rho) = \{f : [\rho]^{\frac{1}{p(x)}}\} \in L^{p(\cdot)}(\Omega)\},$$

$$\|f\|_{L^{p(\cdot)}(\Omega,\rho)} = \inf\left\{\lambda > 0 : \int\limits_{\Omega} \rho(x)\left|\frac{f(x)}{\lambda}\right|^{p(x)} dx \leq 1\right\},$$

where $\rho(x) \geq 0$ a.e. and $|\{x \in \Omega : \rho(x) = 0\}|$.

From Holder inequality for the $L^{p(\cdot)}(\Omega)$-spaces we have

$$|\int\limits_{\Omega} u(x)v(x)dx| < (\frac{1}{p^-} + \frac{1}{(p')^-})\|u\|_{L^{p'(\cdot)}(\Omega)}\|v\|_{L^{p(\cdot)}(\Omega)}.$$

2.2 The Maximal Operator in $L^{p(\cdot)}$

For a locally integrable function f on R^n, the Hardy-Littlewood maximal operator M is defined by

$$Mf(x) = \sup_{\delta>0} \frac{1}{|B(x,\delta)|} \int\limits_{B(x,\delta)} |f(y)|dy,$$

where $B(x, \delta)$ denotes the open ball centered at $x \in R^n$ and radius $\delta > 0$. The sufficiency of condition (1), provided by the next theorem, was proved by Diening [5].

Theorem 1 (Diening) *Let $\Omega \in R^n$ be an open, bounded domain, and let $p : \Omega \to [1, \infty)$ satisfy (1) and be such shat $1 < p^- \leq p(x) \leq p^+ < \infty$. Then the maximal operator is bounded on $L^{p(\cdot)}(\Omega)$:*

$$\|Mf\|_{p(x),\Omega} \leq C(p(x), \Omega)\|f\|_{p(x),\Omega}.$$

2.3   The Riesz Potential Operator on Variable Lebesgue Spaces

The boundedness the Riesz potential operator on variable Lebesgue spaces was first considered in [22], where the Sobolev type theorem for bounded domains was proved under the assumption that the maximal operator is bounded in $L^{p(\cdot)}$ from $L^{q(\cdot)}$. L. Diening in [6] proved the Sobolev theorem on for satisfying the local logarithmic condition (2) and constant at infinity.

Theorem 2 ([6]) *Let $0 < \alpha < n$ and let $1 < p^- \leq p(x) \leq p^+ < \frac{n}{\alpha}$. Assume also that p satisfies the log-Holder conditions (1) and (2), then there exists $C > 0$ such that*

$$\|I^\alpha f\|_{L^{q(\cdot)}(R^n)} \leq C \|f\|_{L^{p(\cdot)}(R^n)},$$

where $q(x)$ is the Sobolev exponent given by $\frac{1}{q(x)} = \frac{1}{p(x)} - \frac{\alpha}{n}, x \in R^n.$

Lemma 1 ([23]) *Let $e_1 = (1, 0, ..., 0)$. For the integral*

$$J_{a,b}(t) = \int\limits_{|y|<t} \frac{dy}{|y|^a |y - e_1|^b}, 0 < t < \infty,$$

where $a < n, b < n, a + b < n$, the following estimate is valid

$$J_{a,b}(t) \leq C \frac{6^{|a|+|b|} \, t^{n-a}}{(n-a)(n-b)(n-a-b)(1+t)^b}, 0 < t < \infty,$$

where $C > 0$ is an absolute constant not depending on t, a and (depending oily on n.)

3   Unilateral Ball Potentials

Series of problem in mathematical physics (see, for example, [2]) are reduced to the reversion of the following integral operators,

$$I_\Omega^\alpha \varphi = c_{n,\alpha} \int\limits_\Omega \frac{\varphi(y)}{|x - y|^{n-\alpha}} dy, x \in \Omega \subset R^n, \quad 0 < \alpha < n, \tag{3}$$

named by Riesz potentials.

Let Ω be a full-sphere in R^n or area supplementing a full-sphere to only R^n. In the given work it is considered one-sided spherical potentials in Lebesgue spaces $L^{p(\cdot)}(\Omega)$ with variable exponent $p(x)$.

Lemma 2 ([19]) *At* $0 < \alpha < n$, $|x| \neq |y|$ *relations hold*

$$\frac{c_{n,\alpha}}{|x-y|^{n-\alpha}} = \frac{2^{n-\alpha}}{\Gamma^2(\frac{\alpha}{2})\omega_{n-1}^2} \int\limits_{|z|<\min(|x|,|y|)} \frac{(|x|^2-|z|^2)^{\alpha/2}(|y|^2-|z|^2)^{\alpha/2}}{|x-z|^n|y-z|^n|z|^\alpha} dz, \quad (4)$$

$$\frac{c_{n,\alpha}}{|x-y|^{n-\alpha}} = \frac{2^{n-\alpha}}{\Gamma^2(\frac{\alpha}{2})\omega_{n-1}^2} \int\limits_{|z|>\max(|x|,|y|)} \frac{(|z|^2-|x|^2)^{\alpha/2}(|z|^2-|y|^2)^{\alpha/2}}{|x-z|^n|y-z|^n|z|^\alpha} dz. \quad (5)$$

Let's consider Riesz potential

$$I^\alpha\varphi = c_{n,\alpha} \int\limits_{R^n} \frac{\varphi(y)}{|x-y|^{n-\alpha}} dy, \quad \varphi \in \mathcal{L}_p(R^n), 1 \leq p < \frac{n}{\alpha}. \quad (6)$$

Substituting in (6) kernel under formulas (4), (5) and changing the integration order, we obtain the following [18–20, 25]

$$I^\alpha\varphi = 2^{-\alpha}B_+^{\frac{\alpha}{2}}|y|^{-\alpha}B_-^{\frac{\alpha}{2}}\varphi, \quad I^\alpha\varphi = 2^{-\alpha}B_-^{\frac{\alpha}{2}}|y|^{-\alpha}B_+^{\frac{\alpha}{2}}\varphi, \quad (7)$$

where

$$B_+^{\frac{\alpha}{2}}\varphi = \frac{2}{\Gamma(\frac{\alpha}{2})\omega_{n-1}} \int\limits_{|y|<|x|} \frac{(|x|^2-|y|^2)^{\alpha/2}}{|x-y|^n}\varphi(y)dy, \quad (8)$$

$$B_-^{\frac{\alpha}{2}}\varphi = \frac{2}{\Gamma(\frac{\alpha}{2})\omega_{n-1}} \int\limits_{|y|>|x|} \frac{(|y|^2-|x|^2)^{\alpha/2}}{|x-y|^n}\varphi(y)dy. \quad (9)$$

Potential operator $I^\alpha\varphi$ in formulas (7) is called B factorization. Factorization (7) is convenient in studying potentials (3) in case when the region Ω is a sphere. Integral operators in (8), (9) are called unilaterals ball potentials.

Unilaterals ball potentials of order $\alpha > 0$ in a spherical layer $U(a,b), 0 \leq a < b \leq \infty$, are defined by:

$$B_{a+}^\alpha\varphi = \frac{2}{\Gamma(\alpha)\omega_{n-1}} \int\limits_{a<|y|<|x|} \frac{(|x|^2-|y|^2)^\alpha}{|x-y|^n}\varphi(y)dz,$$

$$B_{b-}^\alpha\varphi = \frac{2}{\Gamma(\alpha)\omega_{n-1}} \int\limits_{|x|<|y|<b} \frac{(|y|^2-|x|^2)^\alpha}{|x-y|^n}\varphi(y)dz.$$

Potentials $B_{a+}^\alpha\varphi$ are named left sided, and $B_{b-}^\alpha\varphi$ right sided. At $a=0, b=\infty$ we will write accordingly $B_+^\alpha\varphi$, $B_-^\alpha\varphi$.

4   The Main Statement

For $0 \leq \alpha < \frac{n}{p^+}$ and $p \in P(R^n)$ we define $q \in P(R^n)$ point-wise by $\frac{1}{q(x)} := \frac{1}{p(x)} - \frac{\alpha}{n}$ for all $x \in R^n$. Then $q \in P(R^n)$ with

$$1 < \frac{np^-}{n - \alpha p^-} = (q)^- \leq (q)^+ = \frac{np^+}{n - \alpha p^+} < \infty.$$

It is clear that $q \in P^{log}(R^n)$ with $c_{log}(p) = c_{log}(p)$ if $p \in P^{log}(R^n)$.

Lemma 3 *Let $\Omega \subset R^n$—be a bounded open set and $0 < \alpha < n$. Let $p \in P^{log}(\Omega)$ with $1 < p^- \leq p^+ < \frac{n}{\alpha}$. If $k \geq max\{\frac{p^+}{n-\alpha p^+}, 1\}$, then*

$$|B_+^\alpha f| \leq C|x|^\alpha k^{\frac{1}{p^+}} Mf(x)^{1 - \frac{\alpha p(x)}{n}} \tag{10}$$

for all $f \in L^{p(\cdot)}(\Omega)$ with $\|f\|_{p(\cdot)} \leq 1$ and every $x \in \Omega \subset R^n$. The constant depends only on $\alpha, n, c_{log}(p)$ and $diam(\Omega)$.

Proof To estimate $B_+^\alpha f$ we observe that $|x| - |y| \leq |x - y|$ and $|x| + |y| \leq 2|x|$, so that

$$|B_+^\alpha f| \leq 2^\alpha \gamma_{n,\alpha}|x|^\alpha \int_{R^n} \frac{1}{|x - y|^{n-\alpha}} |f(y)|dy = E(x, \delta) + F(x, \delta), \tag{11}$$

where

$$E(x, \delta) := 2^\alpha \gamma_{n,\alpha}|x|^\alpha \int_{|y-x|<\delta} \frac{1}{|x - y|^{n-\alpha}} |f(y)|dy,$$

$$F(x, \delta) := 2^\alpha \gamma_{n,\alpha}|x|^\alpha \int_{|y-x|>\delta} \frac{1}{|x - y|^{n-\alpha}} |f(y)|dy.$$

Therefore,

$$E(x, \delta) = 2^\alpha \gamma_{n,\alpha}|x|^\alpha \int_{|y-x|<\delta} \frac{1}{|x - y|^{n-\alpha}} |f(y)|dy =$$

$$2^\alpha \gamma_{n,\alpha}|x|^\alpha \sum_{k=1}^\infty \int_{2^{-k}\delta<|y-x|<2^{-k+1}\delta} \frac{1}{|x - y|^{n-\alpha}} |f(y)|dy \leq$$

$$2^\alpha \gamma_{n,\alpha}|x|^\alpha \sum_{k=1}^\infty 2^{-k\alpha+kn}\delta^{\alpha-n} \frac{(2^{-k+1}\delta)^n}{(2^{-k+1}\delta)^n} \int_{2^{-k}\delta<|y-x|<2^{-k+1}\delta} |f(y)|dy \leq$$

$$2^{n+\alpha} \gamma_{n,\alpha} |x|^\alpha \delta^\alpha \sum_{k=1}^{\infty} 2^{-k\alpha} M f(x). \tag{12}$$

By (12), we have

$$E(x, \delta) \leq \frac{2^n \gamma_{n,\alpha} |x|^\alpha \delta^\alpha}{2^\alpha - 1} M f(x) \leq C |x|^\alpha \delta^\alpha M f(x) \tag{13}$$

with same absolute constant $C = C(\alpha_0, n) > 0$ not depending on x and δ.

The second function $F(x, \delta)$ is written as follows

$$F(x, \delta) = 2^\alpha \gamma_{n,\alpha} |x|^\alpha \int\limits_{R^n \setminus B(x,\delta)} \frac{1}{|x - y|^{n-\alpha}} |f(y)| dy.$$

Set $B := B(x, \delta)$. Using Holders inequality and taking into account that the fact $\|f\|_{p(\cdot)} \leq 1$, we have

$$F(x, \delta) \leq 2^\alpha \gamma_{n,\alpha} |x|^\alpha \|f\|_{p(\cdot)} \|\chi_{R^n \setminus B} |x - \cdot|^{\alpha-n} \|_{p'(\cdot)} \leq$$

$$2^\alpha \gamma_{n,\alpha} |x|^\alpha \|\chi_{R^n \setminus B} |x - \cdot|^{\alpha-n} \|_{p'(\cdot)} = 2^\alpha \gamma_{n,\alpha} |x|^\alpha \|\chi_{R^n \setminus B} |x - \cdot|^{-n} \|_{s(\cdot)}^{\frac{n-\alpha}{n}},$$

where $s := \frac{n-\alpha}{n} p'$. Next we note that

$$M(\chi_B |B|^{-1})(y) \geq \frac{1}{|2|x - y||^n} \int\limits_{B(y,2|y-x|)} \chi_B(z) dz = |B(y, 2|y - x|)|^{-1} = C |y - x|^{-n}$$

for all $y \in R^n \setminus B(x, \delta)$. Therefore,

$$C \chi_{R^n \setminus B(x,\delta)} |y - x|^{-n} \leq M(\chi_B |B|^{-1})(y)$$

for all $y \in R^n$. Combining the previous estimates, we find that

$$F(x, \delta) \leq 2^\alpha \gamma_{n,\alpha} |x|^\alpha |B|^{\frac{\alpha-n}{n}} \|M(\chi_B)\|_{s(\cdot)}^{\frac{n-\alpha}{n}} \leq 2^\alpha \gamma_{n,\alpha} |x|^\alpha ((s)^-)'|^{\frac{\alpha-n}{n}} \|\chi_B\|_{s(\cdot)}^{\frac{n-\alpha}{n}},$$

where we used Theorem 1 for the boundedness of M on $L^{q(\cdot)}(R^n)$. In addition, we conclude that

$$\|\chi_B\|_{s(\cdot)}^{\frac{n-\alpha}{n}} = \|\chi_B\|_{p'(\cdot)} \leq C |B|^{\frac{1}{p'_B}},$$

where the second estimate follows from Corollary 4.5.9 in [9]. Combining these estimates, we obtain

$$F(x, \delta) \leq 2^{\alpha} \gamma_{n,\alpha} |x|^{\alpha} |((s)^-)'|^{\frac{\alpha-n}{n}} |B|^{\frac{\alpha-n}{n} + \frac{1}{p_B'}} \leq 2^{\alpha} C |x|^{\alpha} (\frac{n-\alpha}{n-\alpha p^+})^{\frac{n-\alpha}{n}} |B|^{\frac{1}{q_B}}. \quad (14)$$

Let $x \in \Omega$ and let $0 < \delta < 2diam\Omega$ be a number to be specified later. Having substituted (13) and (14) into (11), we have

$$|B_+^{\alpha} f| \leq C|x|^{\alpha} [\delta^{\alpha} Mf(x) + (\frac{n-\alpha}{n-\alpha p^+})^{\frac{n-\alpha}{n}} |B|^{\frac{1}{q_B}}] \leq$$

$$C|x|^{\alpha} [\delta^{\alpha} Mf(x) + (k)^{\frac{1}{(p+)'}} \delta^{\frac{-n}{q_{B(x,\delta)}}}]. \quad (15)$$

Since, $0 < \delta < 2diam\Omega$, we have $\delta^{\frac{-n}{q_{B(x,\delta)}}} \approx \delta^{\frac{-n}{q(x)}}$.

If $[Mf(x)]^{\frac{-p(x)}{n}} < 2diam\Omega$, we choose $\delta = [Mf(x)]^{\frac{-p(x)}{n}}$. Then estimate (15) gives (10). On the other hand, if $\delta = [Mf(x)]^{\frac{-p(x)}{n}} > 2diam\Omega$, we choose $\delta = 2diam\Omega$. Now we have $\delta^{\alpha} = [Mf(x)]^{\frac{-p(x)\alpha}{n}}$, so (10) follows directly from Lemma 6.1.4 in [9].

Lemma 4 *Let $p \in P^{log}(R^n), 0 < \alpha < n$, and $1 < p^- \leq p^+ < \frac{n}{\alpha}$. Then for any $m > n$ there exists $C > 0$ only depending on $c_{log}(p), p^+, \alpha$, and n such that the following estimate*

$$||x|^{-\alpha} B_+^{\alpha}|^{q(x)} \leq |I^{\alpha} f|^{q(x)} \leq CMf(x)^{p(x)} + h(x)$$

holds for all $f \in L^{p(\cdot)}$ with $\|f\|_{p(\cdot)} \leq 1$ and all $x \in R^n$, where $h \in L^1(R^n) \bigcap L^{\infty} (R^n), \frac{1}{q(x)} := \frac{1}{p(x)} - \frac{\alpha}{n}$.

The proof of Lemma 4 is similar to the proof of Lemma 6.1.8 in [9].

Theorem 3 *Let $p \in P^{log}(R^n), 0 < \alpha < n$ and $1 < p^- \leq p^+ < \frac{n}{\alpha}$. Then the following estimate*

$$\||x|^{-\alpha} B_+^{\alpha} f\|_{q(.)} \leq C\|f\|_{p(.)},$$

holds for all $f \in L^{p(\cdot)}$ with $\|f\|_{p(\cdot)} \leq 1$, where the constant C depends on p only via $c_{log}(p), p^-$ and $p^+, \frac{1}{q(x)} := \frac{1}{p(x)} - \frac{\alpha}{n}$.

Proof Let $h \in L^1(R^n) \bigcap L^{\infty}(R^n)$ be as in Lemma 4. Let $\|f\|_{p(\cdot)} \leq 1$ and thus $A_{p(\cdot)}(f) \leq 1$ by the unit ball property. Integrating in all sides of the estimate in Lemma 4 over $x \in R^n$ yields

$$A_{q(\cdot)}(|x|^{-\alpha} B_+^{\alpha} f) \leq A_{q(\cdot)}(I^{\alpha} f) \leq A_{p(\cdot)}(Mf) + A_1(h) \leq A_{p(\cdot)}(Mf) + c.$$

By Theorem 1 we have that, M is bounded on $L^{p(\cdot)}(R^n)$ and so $A_{p(\cdot)}(Mf) \leq 1$ implies $A_{p(\cdot)}(Mf) \leq C$, with $q^+ < \infty$. Hence, $A_{q(\cdot)}(|x|^{-\alpha} B_+^{\alpha} f) \leq A_{q(\cdot)}(I^{\alpha} f) \leq C$ and therefore, $\||x|^{-\alpha} B_+^{\alpha} f\|_{q(.)} \leq C$, with $q^+ < \infty$. Since, I^{α} is sublinear, a scaling argument completes the proof.

Theorem 4 *Let $p \in P_1(\Omega)$, $0 < \alpha < n$ and $1 < p^- \leq p^+ < \infty$. If $p^- > \frac{n}{\alpha}$, then*

$$|B_+^\alpha f| \leq C|x|^{\frac{2\alpha-n}{P(x)}+n}, \tag{16}$$

for all $f \in L^{p(\cdot)}(R^n)$ with $\|f\|_{p(\cdot)} \leq 1$ and all $x \in R^n$, where the constant C depends on α, n, $c_{log}(p)$.

Proof For points $x, y \in R^n$ let us denote $r = |x|$, $\rho = |y|$, $x' = \frac{x}{|x|}$, $y' = \frac{y}{|y|}$. Applying Holders Inequality, we get

$$|B_+^\alpha f| \leq \|\varphi(x, y)\|_{p'(\cdot)} \|f\|_{p(\cdot)},$$

where

$$\varphi(x, y) := \gamma_{n,\alpha} \frac{(|x|^2 - |y|^2)^\alpha}{|x - y|^n} \chi_{B(0,|x|)}(y).$$

We use the property

$$\rho_{p(\cdot)}\left(\frac{f}{b}\right) \leq a \Rightarrow \|f\|_{p(\cdot)} \leq ab^\nu,$$

where

$$\nu = \begin{cases} \frac{1}{p^-}, b \geq 1, \\ \frac{1}{p^+}, \quad b < 1. \end{cases}$$

We can prove that

$$\rho_{p'(\cdot)}\left(\frac{\varphi(x, \cdot)}{|x|^{2\alpha-n}}\right) \leq C|x|^n, \tag{17}$$

which implies that

$$\|\varphi(x, \cdot)\|_{p'(\cdot)} \leq C|x|^{\frac{2\alpha-n}{P(x)}+n},$$

where

$$P(x) = \begin{cases} (p')^-, |x| \geq 1, \\ (p')^+, \quad |x| \leq 1. \end{cases}$$

Let us try to prove (19). We have

$$\rho_{p'(\cdot)}\left(\frac{\varphi(x, \cdot)}{|x|^{2\alpha-n}}\right) = \gamma_{n,\alpha} \int\limits_{|y|<|x|} \frac{(|x|^2 - |y|^2)^{\alpha p'(\cdot)}}{|x|^{(2\alpha-n)p'(\cdot)}|x - y|^{np'(\cdot)}} dy \leq$$

$$\gamma_{n,\alpha} 2^{\alpha(p')+} \int\limits_{|y|<|x|} \frac{dy}{\left|x' - \frac{y}{|x|}\right|^{(n-\alpha)p'(\cdot)}}$$

Taking $y = |x|rot_x\xi$, $rot_x e_1 = x'$, $e_1 = (1, 0, ..., 0)$, we have

$$\rho_{p'(\cdot)}\left(\frac{\varphi(x,\cdot)}{|x|^{2\alpha-n}}\right) \le C|x|^n \int\limits_{|\xi|<1} \frac{d\xi}{|e_1 - \xi|^{(n-\alpha)p'}(|x|rot_x\xi)}$$

$$= C|x|^n \left(\int\limits_{|\xi|<1,|e_1-\xi|<1} + \int\limits_{|\xi|<1,|e_1-\xi|>1} \right) \frac{d\xi}{|e_1 - \xi|^{(n-\alpha)p'}(|x|rot_x\xi)}$$

$$\le C|x|^n \left(\int\limits_{|\xi|<1,|e_1-\xi|<1} \frac{d\xi}{|e_1 - \xi|^{(n-\alpha)(p')^+}} + \int\limits_{|\xi|<1,|e_1-\xi|>1} \frac{d\xi}{|e_1 - \xi|^{(n-\alpha)(p')^-}} \right)$$

Here, we obtain following inequality based on Lemma 1

$$\rho_{p'(\cdot)}\left(\frac{\varphi(x,\cdot)}{|x|^{2\alpha-n}}\right) \le C|x|^n \left(\frac{3^{(n-\alpha)(p')^+}}{n(n-(n-\alpha)(p')^+)^2} + \frac{3^{(n-\alpha)(p')^-}}{n(n-(n-\alpha)(p')^-)^2}\right). \quad (18)$$

Using (18) and (17), we have (16). This completes the proof.

Theorem 5 Let $p \in P(\Omega)$, $0 < \alpha < n$ and $1 < p^- \le p(x) \le p^+ < \infty$. If $p^- > \frac{n}{\alpha}$, then

$$\||x|^{\frac{n-2\alpha}{P(x)}-n} B_+^\alpha f\|_{L^{p(\cdot)}(\Omega)} \le C\|f\|_{L^{p(\cdot)}(\Omega)},$$

for all $f \in L^{p(\cdot)}(\Omega)$ with $\|f\|_{L^{p(\cdot)}(\Omega)} \le 1$ and all $x \in \Omega$, where the constant depends on α and n.

The proof of this theorem is similar to the proof of the Theorem 3.

Theorem 6 Let $\Omega \subset R^n$ be a bounded domain in $0 < \alpha < n$ and let $p \in P^{log}(R^n)$ with $1 < p^- \le p^+ < \frac{n}{\alpha}$. Then the following estimate

$$\|B_-^\alpha f\|_{L^{q(\cdot)}(\Omega)} \le C\|f\|_{L^{p(\cdot)}(\Omega)}, \quad (19)$$

holds for all $f \in L^{p(\cdot)}(\Omega)$ with $\||y|^\alpha f\|_{p(\cdot)} \le 1$, where the constant C depends on α, n, and $c_{log}(p)$, $\frac{1}{q(x)} := \frac{1}{p(x)} - \frac{\alpha}{n}$.

Proof In the integral $|B_-^\alpha f|$, we observe that $|x| - |y| \le |x - y|$ and $|x| + |y| \le 2|y|$, so that

$$|B_-^\alpha f| \le 2^\alpha \gamma_{n,\alpha} \int\limits_{|y|>|x|} \frac{|y|^\alpha}{|x-y|^{n-\alpha}} |f(y)|dy \le 2^\alpha \gamma_{n,\alpha} \int\limits_{R^n} \frac{|y|^\alpha}{|x-y|^{n-\alpha}} |f(y)|dy =$$

$$E_1(x, \delta) + F_1(x, \delta), \quad (20)$$

where

$$E_1(x, \delta) = 2^\alpha \gamma_{n,\alpha} \int\limits_{|y-x|<\delta} \frac{|y|^\alpha}{|x - y|^{n-\alpha}} |f(y)| dy,$$

$$F_1(x, \delta) = 2^\alpha \gamma_{n,\alpha} \int\limits_{|y-x|>\delta} \frac{|y|^\alpha}{|x - y|^{n-\alpha}} |f(y)| dy.$$

For $E_1(x, \delta)$, via the standard binary decomposition

$$E_1(x, \delta) \le 2^\alpha \gamma_{n,\alpha} \sum_{k=1}^\infty \int\limits_{2^{-k}\delta<|y-x|<2^{-k+1}\delta} \frac{1}{|x - y|^{n-\alpha}} |y|^\alpha |f(y)| dy$$

we obtain

$$E_1(x, \delta) \le \frac{2^{n+\alpha} \gamma_{n,\alpha} \delta^\alpha}{2^\alpha - 1} M g(x) \le C \delta^\alpha M g(x), \tag{21}$$

where $g := ||y|^\alpha f|$, with some absolute constant $C_1 = C_1(\alpha, n) > 0$ not depending on x and δ. We assume that $\| |y|^\alpha f \|_{p(\cdot)} \le 1$. For the term we use the Holder inequality and obtain

$$F_1(x, \delta) \le 2^\alpha \gamma_{n,\alpha} \| |y|^\alpha f \|_{p(\cdot)} \| \chi_{\Omega \setminus B} |x - \cdot|^{\alpha-n} \|_{p'(\cdot)}. \tag{22}$$

We can apply Theorem 1.17 in [22], its assumptions being satisfied due to conditions of our theorem. Then we have

$$\| \chi_{\Omega \setminus B} |x - \cdot|^{\alpha-n} \|_{p'(\cdot)} \le C \delta^{-\frac{n}{q(x)}}. \tag{23}$$

From (20), in view of (21)–(23), we obtain

$$|B_-^\alpha f| \le C(\delta^\alpha M(|y|^\alpha f) + \delta^{-\frac{n}{q(x)}}).$$

Minimizing with respect to δ, at $\delta = (M(|y|^\alpha f))^{-\frac{p(x)}{n}}$ we get

$$|B_-^\alpha f| \le C M(|y|^\alpha f)^{\frac{p(x)}{q(x)}}.$$

Hence,

$$\int_\Omega |B_-^\alpha f|^{q(x)} dx \le C \int_\Omega |(M(|y|^\alpha f))|^{p(x)} dx \le C. \tag{24}$$

Then $\int_\Omega |(M(|y|^\alpha f))|^{p(x)} dx \le C$ and by (24) we obtain that

$$\int_\Omega |B_-^\alpha f|^{q(x)} dx \le C$$

for all $f \in L^{p(\cdot)}(\Omega)$ with $\| |y|^\alpha f \|_{p(\cdot)} \le 1$, which is equivalent to (19).

References

1. Almeida, Samko, S.G.: Characterization of Riesz and Bessel potentials on variable Lebesgue spaces. J. Funct. Spaces Appl. **4**(2), 113–144 (2006)
2. Copson, E.T.: On the problem of the electrified disk. Proc. Edinburd Math. Soc. **8**, 14–19 (1947)
3. Cruz-Uribe, D., Fiorenza, A., Neugebauer, C.J.: The maximal function on variable $L^{p(\cdot)}$ spaces. Ann. Acad. Sci. Fenn. Math. **28**, 223–238 (2003)
4. Cruz-Uribe, D., Fiorenza, A., Martell, J.M., Perez, C.: The boundedness of classical operators on variable $L^{p(x)}$ spaces. Ann. Acad. Sci. Fenn. Math. **31**, 239–264 (2006)
5. Diening, L.: Maximal function on generalized Lebesgue spaces $L^{p(\cdot)}$. Math. Inequal. Appl. **7**(2), 245–253 (2004)
6. Diening, L.: Riesz potential and Sobolev embeddings on generalized Lebesgue and Sobolev spaces. Math. Nachr. **268**, 31–43 (2004)
7. Diening, L.: Maximal function on Musielak-Orliz spaces and generalized Lebesgue spaces, Preprint (2004)
8. Diening, L., RUziCka, M.: Calderon-Zygmund operators on generalized Lebesgue spaces $L^{p(x)}$ and problems related to fluid dynamics. Preprint Mathematische Fakultat, Albert-Ludwigs-Universitat Freiburg (21/2002, 04.07.2002), 1–20 (2002)
9. Diening, L., Harjulehto, P., Hasto, P., Ruzicka, M.: Lebesgue and Sobolev Spaces with Variable Exponents. Spring Lectute Notes, vol. 2017. Springer, Berlin (2011)
10. Edmunds, D.E., Kokilashvili, V., Meskhi, A.: One-sided operators in $L^{p(x)}$ spaces. Math. Nachr. **281**(11), 1525–1548 (2008)
11. Fan, X., Zhao, D.: On the spaces $L^{p(x)}$ and $W^{k,p(x)}$. J. Math. Anal. Appl. **263**(2), 424–446 (2001)
12. Kokilashvili, V.: On a progress in the theory of integral operators in weighted Banach function spaces. In: Drabek, P., Rakosnik, J. (eds.) FSDONA 2004 Proceedings, Milovy, Czech Republic, 2004, pp. 152–175. Math. Inst. Acad. Sci. Czech Rep, Prague (2005)
13. Kokilashvili, V., Samko, S.: Maximal and fractional operators in weighted $L^{p(x)}$ - spaces. Proc. A. Razmadze Math Inst. **129**, 145–149 (2002)
14. Kokilashvili, V., Samko, N., Samko, S.: The maximal operator in weighted variable spaces $L^{p(\cdot)}$. J. Funct. Spaces Appl. **5**, 299–317 (2007)
15. Kovacik, O., Rakosnik, J.: On spaces $L^{p(x)}$ and $W^{k,p(x)}$. Czechoslovak Math. J. **41**(116), 592–618 (1991)
16. Nekvinda, A.: Hardy-Littlewood maximal operator on $L^{p(x)}(R^n)$. Math. Inequal. Appl. **7**(2), 255–265 (2004)
17. Rafeiro, H., Yakhshiboev, M.: The Chen-Marchaud fractional integro-differentiation in the variable exponent Lebesgue spaces. Fract. Calc. Appl. Anal. **3**, 501–518 (2011)
18. Rubin, B.S.: With. Unilateral spherical potentials and address potentials Rissa on - to a measured sphere of its venshnost. Deponierted in VINITI, No. 5150-84, Dep. (1984)
19. Rubin, B.S.: Fractional integrals and weakly singular integral equations ofthe first kind in the n-dimensional ball. J. Anal. Math. **63**, 55–102 (1994)
20. Rubin, B.S.: Fractional Integrals and Potentials. Addison Wesley Longman, Essex (1996)
21. Ruzicka, M.: Electroreological Fluids: Modeling and Mathematical Theory. Springer, Lecture Notes in Math., vol. 1748, 176 p (2000)
22. Samko, S.G.: Convolution and potential type operators in $L^{p(x)}(R^n)$. Integral Transform. Spec. Funct. **7**(3–4), 261–284 (1998)
23. Samko, S.G.: Hardy-Littlewood-Stein-Weiss inequality in the Lebesgue spaces with variable exponent. Fract. Calc. Calc. Appl. Anal. **6**(4), 421–440 (2003)
24. Samko, S.G.: On a progress in the theory of Lebesgue spaces with variable exponent: maximal and singular operators. Integral Transforms Spec. Funct. **16**, 461–482 (2005)
25. Samko, S.G., Yakhshiboev, M.U.: Connections between unilateral ball potentials via operators that are singular with respect to the radial variable, Deponierted in VINITI, 172-B92 January 16 (1992)

26. Samko, S.G., Kilbas, A.A., Marichev, O.I.: Fractional integrals and derivatives. Theory and applications. Gordon and Breach Seience Publishers, London (1993)
27. Sankar, T.S., Fabrikant, V.I.: Investigations of a two-dimensional integral equation in the theory of elasticity and electrostatics. J. mes. Theor et appl. **2**(2), 285–299 (1983)
28. Sharapudinov, I.I.: The topology of the space $L^{p(t)}([0, 1])$ (in Russian). Mat. Zametkai **26**(4), 613–632 (1979)
29. Zhikov, V.V.: On some variational problems. Russian J. Math. Phys. **5**(1), 105–116 (1997)
30. Zhikov, V.V., Pastukhova, S.E.: On the improved integrability of the gradient of solutions of elliptic equations with a variable nonlinearity exponent. Mat. Sb. **199**(12), 19–52 (2008)

Printed in the United States
By Bookmasters